Just The facts101

Textbook Key Facts

e-Study Guide

by **cram101**
Textbook NOT Included

Textbook Outlines, Highlights, and Practice Quizzes

Signals and Systems

by Oppenheim & Willsky, 2nd Edition

All "Just the Facts101" Material Written or Prepared by Cram101 Publishing

Title Page

WHY STOP HERE... THERE'S MORE ONLINE

With technology and experience, we've developed tools that make studying easier and efficient. Like this Craml0l textbook notebook, Craml0l.com offers you the highlights from every chapter of your actual textbook. However, unlike this notebook, Craml0l.com gives you practice tests for each of the chapters. You also get access to in-depth reference material for writing essays and papers.

By purchasing this book, you get 50% off the normal subscription free!. Just enter the promotional code **'DK73DW2097'** on the Cram101.com registration screen.

CRAMI0I.COM FEATURES:

Outlines & Highlights
Just like the ones in this notebook, but with links to additional information.

Integrated Note Taking
Add your class notes to the Cram101 notes, print them and maximize your study time.

Problem Solving
Step-by-step walk throughs for math, stats and other disciplines.

Practice Exams
Five different test taking formats for every chapter.

Easy Access
Study any of your books, on any computer, anywhere.

Unlimited Textbooks
All the features above for virtually all your textbooks, just add them to your account at no additional cost.

Be sure to use the promo code above when registering on Craml0l.com to get 50% off your membership fees.

STUDYING MADE EASY

This Cram101 notebook is designed to make studying easier and increase your comprehension of the textbook material. Instead of starting with a blank notebook and trying to write down everything discussed in class lectures, you can use this Cram101 textbook notebook and annotate your notes along with the lecture.

Our goal is to give you the best tools for success.

For a supreme understanding of the course, pair your notebook with our online tools. Should you decide you prefer Cram101.com as your study tool,

we'd like to offer you a trade...

Our Trade In program is a simple way for us to keep our promise and provide you the best studying tools, regardless of where you purchased your Cram101 textbook notebook. As long as your notebook is in *Like New Condition**, you can send it back to us and we will immediately give you a Cram101.com account free for 120 days!

Let The *Trade In* Begin!

THREE SIMPLE STEPS TO TRADE:

1. Go to www.cram101.com/tradein and fill out the packing slip information.
2. Submit and print the packing slip and mail it in with your Cram101 textbook notebook.
3. Activate your account after you receive your email confirmation.

* Books must be returned in *Like New Condition*, meaning there is no damage to the book including, but not limited to; ripped or torn pages, markings or writing on pages, or folded / creased pages. Upon receiving the book, Cram101 will inspect it and reserves the right to terminate your free Cram101.com account and return your textbook notebook at the owners expense.

"Just the Facts101" is a Cram101 publication and tool designed to give you all the facts from your textbooks. Visit Cram101.com for the full practice test for each of your chapters for virtually any of your textbooks.

Cram101 has built custom study tools specific to your textbook. We provide all of the factual testable information and unlike traditional study guides, we will never send you back to your textbook for more information.

YOU WILL NEVER HAVE TO HIGHLIGHT A BOOK AGAIN!

Cram101 StudyGuides
All of the information in this StudyGuide is written specifically for your textbook. We include the key terms, places, people, and concepts... the information you can expect on your next exam!

Want to take a practice test?
Throughout each chapter of this StudyGuide you will find links to cram101.com where you can select specific chapters to take a complete test on, or you can subscribe and get practice tests for up to 12 of your textbooks, along with other exclusive cram101.com tools like problem solving labs and reference libraries.

Cram101.com
Only cram101.com gives you the outlines, highlights, and PRACTICE TESTS specific to your textbook. Cram101.com is an online application where you'll discover study tools designed to make the most of your limited study time.

By purchasing this book, you get 50% off the normal subscription free!. Just enter the promotional code **'DK73DW2097'** on the Cram101.com registration screen.

www.Cram101.com

Copyright © 2012 by Cram101, Inc. All rights reserved.
"Just the FACTS101"®, "Cram101"® and "Never Highlight a Book Again!"® are registered trademarks of Cram101, Inc.
ISBN(s): 9781618128935. PUBE-2.2012118

Learning System

Signals and Systems
Oppenheim & Willsky, 2nd

CONTENTS

1. SIGNALS AND SYSTEMS 14
2. LINEAR TIME-INVARIANT SYSTEMS 25
3. FOURIER SERIES REPRESENTATION OF PERIODIC SIGNALS 38
4. THE CONTINUOUS - TIME FOURIER TRANSFORM 48
5. THE DISCRETE-TIME FOURIER TRANSFORM 57
6. TIME AND FREQUENCY CHARACTERIZATION OF SIGNALS AND SYSTEMS 70
7. SAMPLING 81
8. COMMUNICATION SYSTEMS 93
9. THE LAPLACE TRANSFORM 103
10. THE Z - TRANSFORM 113
11. LINEAR FEEDBACK SYSTEMS 126

CHAPTER OUTLINE: KEY TERMS, PEOPLE, PLACES, CONCEPTS
Chapter 1
SIGNALS AND SYSTEMS

- Set1
- Capacitor1
- Force1
- Velocity1
- Pressure1
- Voltage1
- Applied force1
- Digital1
- Function1
- Second1
- Sound1
- System1
- Image1
- Independent variable1
- Density1
- Temperature1
- Speed1
- Power1
- Position1

Chapter 1. SIGNALS AND SYSTEMS

- Matter 1
- Energy 1
- Range 1
- Current 1
- Resistor 1
- Resistance 1
- Time interval 1
- Friction 1
- Magnitude 1
- Units 1
- Dynamics 1
- Kinematics 1
- Noise 1
- Axis 1
- Properties 1
- Negative 1
- Radar 1
- Origin 1
- Reflection 1

Chapter 1. SIGNALS AND SYSTEMS

- Compression1
- Periodic1
- Property1
- Period1
- Trigonometry1
- Symmetry1
- Radioactive decay1
- Decay1
- Hertz1
- Simple harmonic motion1
- Complex number1
- Fundamental frequency1
- Frequency1
- Oscillation1
- Amplitude1
- Cell1
- Cycle1
- Impulse1
- Pulse1

Chapter 1. SIGNALS AND SYSTEMS

- Equation 1
- Quantity 1
- Inertia 1
- Singularity 1
- Ohm's law 1
- Proportionality constant 1
- Voltage drop 1
- Net force 1
- Receiver 1
- Rotation 1
- Capacitance 1
- Charge 1
- Kinetic energy 1
- Stable 1
- Diverge 1
- Gravity 1
- Restoring force 1
- Unstable 1
- Terminal velocity 1

Chapter 1. SIGNALS AND SYSTEMS

Mechanics1

Proof1

Phase1

Complex plane1

CHAPTER HIGHLIGHTS: KEY TERMS, PEOPLE, PLACES, CONCEPTS
Chapter 1. SIGNALS AND SYSTEMS

Set1	In mathematics a set is a collection of related objects. The mathematical usage is similar to the ordinary English meaning of the word. The objects that make up a set are called the elements of the set. If a set contains an unlimited number of elements it is an infinite set. Otherwise it is a finite set.
Capacitor1	Electrical device used to store charge and energy in the electrical field is referred to as a capacitor.
Force1	Force refers to agent that results in accelerating or deforming an object.
Velocity1	The ratio of change in position with respect to the time interval over which the change occurred is referred to as velocity.
Pressure1	Force per unit area is referred to as the pressure.
Voltage1	Voltage refers to potential difference. It is a measure of the change in energy that one coulomb of electric charge undergoes when moved between 2 points.
Applied force1	We make a somewhat arbitrary distinction among the forces acting in a dynamical system. The categories are applied force, centering force and drag or friction force. The applied force is taken to be that force which is applied to the moving parts of a system by an outside agent as for example the force applied to a pendulum by someone pushing it or the force applied to a piece of metal by the magnetic field of an electromagnet. An applied force results in an energy transfer across the system boundary.
Digital1	In electronics, meaning 'coded as numbers'. Digital means having discrete values as on a digital display.
Function1	A mathematical function is a rule relating two sets of objects. Here we will restrict ourselves to objects that are numbers or vectors. One of the sets is called the domain of the function, the other is called the range of the function. Functions are frequently expressed as equations as for example Y=X+2. This function is interpreted as follows. For every X in the domain, add 2 to it to get the corresponding Y in the range. Because we are free to choose any X we want, X is called the independent variable. Because once X is chosen Y is fixed, we call Y the dependent variable.
Second1	Second is a SI unit of time.

Chapter 1. SIGNALS AND SYSTEMS

Sound1	A longitudinal wave, usually through the air but can travel through liquids and solids is called sound.
System1	Defined collection of objects is called a system.
Image1	Reproduction of object formed with lenses or mirrors is the image.
Independent variable1	Variable that is manipulated or changed in an experiment is referred to as the independent variable.
Density1	The mass of a given volume of substance. It has units of kg/m3 or g/cm3. When the density is high the particles are closely packed.
Temperature1	Temperature refers to measure of hotness of object on a quantitative scale. In gases, proportional to average kinetic energy of molecules.
Speed1	Speed refers to ratio of distance traveled to time interval.
Power1	Power refers to rate of doing work; rate of energy conversion.
Position1	Separation between object and a reference point is a position.
Matter1	We call the commonly observed particles such as protons, neutrons and electrons matter particles, and their antiparticles, antimatter.
Energy1	Energy refers to non-material property capable of causing changes in matter.
Range1	The range is the set of values that the dependent variable of a function may take on. A range may be finite as in the set of numbers {1, 2, 3.n} or infinite as in all the mumbers between 0 and 1.
Current1	Current refers to a flow of charge. Measured in amps.
Resistor1	Device designed to have a specific resistance is a resistor.
Resistance1	Ratio of potential difference across device to current through it are called the resistance.
Time interval1	Difference in time between two clock readings is referred to as a time interval.

Chapter 1. SIGNALS AND SYSTEMS

Friction1	Force opposing relative motion of two objects are in contact is friction.
Magnitude1	Magnitude refers to the size of a thing, without regard for its sign or direction. Similar to the absolute value of a number but applies to vectors as well.
9. Thus a gev is a billion electron Volts. The mass-energy of a proton or neutron is approximately 1 gev.	The units one uses should be of a size that makes sense for the particular subject at hand. It is easiest to define units in each area of science and then relate them to one another than to go around measuring particle masses in grams or cheese in proton mass units. In particle physics the standard unit is the unit of energy gev. One ev is the amount of energy that an electron gains when it moves through a potential difference of 1 Volt . G stands for Giga, or 10
Dynamics1	Study of motion of particles acted on by forces is called dynamics.
Kinematics1	Study of motion of objects without regard to the causes of this motion is called kinematics.
Noise1	Noise refers to sound consisting of an irregular mixture of frequencies.
Axis1	Axis refers to the imaginary line about which a planet or other object rotates .
Properties1	Properties refers to qualities or attributes that, taken together, are usually unique to an object; for example, color, texture, and size.
Negative1	The sign of the electric charge on the electron is negative.
Radar1	A method of detecting, locating, or tracking an object by using beamed, reflected, and timed radio waves. RADAR also refers to the electronic equipment that uses radio waves to detect, locate, and track objects.
Origin1	Origin refers to the point in a reference frame from which measurements are made. It is the location of the zero value for each axis in the frame.
Reflection1	Reflection refers to the change when light, sound, or other waves bounce backwards off a boundary .
Compression1	In a sound wave, a compression is a region in the material that transmits the wave where the particles are closer together than normal.

Chapter 1. SIGNALS AND SYSTEMS

Periodic1	A dynamical system or function which at some point returns to the same state or value. If a system or function ever revisits the identical state or value it will continue to come back to it again and again in equal intervals of time. That is why we call such a system periodic.
Property1	A characteristic that is inherently associated with the object which is said to have that property. For example the mass of an object is one of its properties. So also might be color, density and many other characteristics. Properties are classified as extensive or intensive. Extensive properties increase in proportion to the size of the object, as mass does for example. Intensive properties are independent of the size of the object. The density for examples remains the same if I cut an object in half and throw half of it away. Things like an object's position or velocity are not considered to be properties of the object. They are not a characteristic of the object only but are also dependent on the reference frame in which the object is located.
Period1	Period refers to time needed to repeat one complete cycle of motion.
Trigonometry1	Trigonometry refers to branch of math that deals with the relationship among angles and sides of triangles.
Symmetry1	Symmetry refers to property that is now charged when operation or reference frame is charged.
Radioactive decay1	Spontaneous change of unstable nuclei into other nuclei is radioactive decay.
Decay1	Any process in which a particle disappears and in its place two or more different particles appear is decay.
Hertz1	Unit of frequency equal to one event or cycle per second is called hertz.
Simple harmonic motion1	Motion caused by linear restoring that has a period independent of amplitude of motion is referred to as simple harmonic motion.
Complex number1	Ordinary numbers that we use for accounting and simple calculations are called real numbers. There is no real number such that the square of that number is -1 since the product of any real number with itself is positive. To remedy this situation, the square root of -1 was defined and given the symbol i. Then a new class of numbers was invented, called imaginary numbers, made up of i times the set of real numbers. A complex number is a number comprised of a real part added to an imaginary part like a+b*i.

Chapter 1. SIGNALS AND SYSTEMS

Fundamental frequency1	Fundamental frequency refers to the lowest frequency that can set up standing waves in an air column or on a string.
Frequency1	Frequency refers to number of occurrences per unit time.
Oscillation1	A to-and-fro or, side-to-side movement is an oscillation.
Amplitude1	In any periodic motion, the maximum displacement from equilibrium is called amplitude.
Cell1	In electricity, a cell is a combination of metals and chemicals that produces a voltage and can cause a current.
Cycle1	In wave motion, one cycle is a trough and a crest for a transverse wave, or a compression and a rarefaction for a longitudinal wave.
Impulse1	Product of force and time interval over which it acts is the impulse.
Pulse1	A wave of short duration confined to a small portion of the medium at any given time is called a pulse. A pulse is also called a wave pulse.
Equation1	An equation is a mathematical expression with an equal sign in it. It signifies that the numerical or vector value on one side of the = is the same as the numerical or vector value on the other side. An equation may include variables and parameters. If any of the variables are rates of change, the equation is called a differential equation.
Quantity1	A numerical value either scalar or vector, which describes some attribute of an object like its position or its velocity. We sometimes speak of physical quantities to signify that we are talking about an object's properties or attributes as opposed to a purely mathematical quantity.
Inertia1	Inertia refers to tendency of object not to change its motion.
Singularity1	The center of a black hole, where the curvature of spacetime is maximal is its singularity. At the singularity, the gravitational tides diverge; no solid object can even theoretically survive hitting the singularity. Although singularities generally predict inconsistencies in theory, singularities within black holes do not necessarily imply that general relativity is incomplete so long as singularities are always surrounded by event horizons. A proper formulation of quantum gravity may well avoid the classical singularity at the centers of black holes.

Chapter 1. SIGNALS AND SYSTEMS

Ohm's law1	Ohm's law refers to the ratio of the potential difference between the ends of a conductor to the current flowing through it is constant; the constant of proportionality is called the resistance, and is different for different materials.
Proportionality constant1	A constant applied to a proportionality statement that transforms the statement into an equation is a proportionality constant.
Voltage drop1	The electric potential difference across a resistor or other part of a circuit that consumes power is referred to as voltage drop.
Net force1	Net force refers to vector sum of forces on an object.
Receiver1	Device that detects electromagnetic waves is called receiver.
Rotation1	Rotation refers to the spin of an object around its central axis. Earth rotates about its axis every 24 hours. A spinning top rotates about its center shaft.
Capacitance1	Ratio of charge stored per increase in potential difference is called capacitance.
Charge1	A property of atomic particles. Electrons and protons have opposite charges and attract each other. The charge on an electron is negative and that on a proton is positive. Measured in coulombs.
Kinetic energy1	Energy of object due to its motion is kinetic energy.
Stable1	Does not decay is stable.
Diverge1	If the dependent variable of a function under iteration increases without limit the function is said to diverge under iteration.
Gravity1	A force with infinite range which acts between objects, such as planets,according to their mass is called gravity.
Restoring force1	The force which tends to bring an oscillating body towards its mean position whenever it is displaced from the mean position is called the restoring force.
Unstable1	Matter that is capable of undergoing spontaneous change, as in a radioactive nuclide or an excited nuclear system. An unstable particle is any elementary particle that spontaneously decays into other particles.

Chapter 1. SIGNALS AND SYSTEMS

Terminal velocity1	Velocity of falling object reached when force of air resistance equals weight is a terminal velocity.
Mechanics1	The study of objects in motion. Mechanics is normally limited to a small number of large slow objects, as opposed to statistical mechanics which deals with large numbers of objects, relativistic mechanics which deals with objects moving near the speed of light and quantum mechanics which deals with objects more or less the size of atoms. Mechanics encompasses the topics of kinematics and dynamics .
Proof1	A measure of ethanol concentration of an alcoholic beverage; proof is double the concentration by volume; for example, 50 percent by volume is 100 proof.
Phase1	The particles in a wave, which are in the same state of vibration, i.e., the same position and the same direction of motion are said to be in the same phase.
Complex plane1	The complex plane contains the set of complex numbers. It is a plane spanned by the set of real numbers, normally along the horizontal axis, and the set of imaginary numbers, normally along the vertical axis.

PRACTICE QUIZ
Chapter 1. SIGNALS AND SYSTEMS

1. _____ refers to ratio of distance traveled to time interval.

    a.
    b.
    c. Speed
    d.

2. Electrical device used to store charge and energy in the electrical field is referred to as a _____.

    a.
    b.
    c. Capacitor
    d.

3. A mathematical _____ is a rule relating two sets of objects. Here we will restrict ourselves to objects that are numbers or vectors. One of the sets is called the domain of the function, the other is called the range of the function. Functions are frequently expressed as equations as for example Y=X+2. This _____ is interpreted as follows. For every X in the domain, add 2 to it to get the corresponding Y in the range. Because we are free to choose any X we want, X is called the independent variable. Because once X is chosen Y is fixed, we call Y the dependent variable.

    a.
    b. Function
    c.
    d.

4. Reproduction of object formed with lenses or mirrors is the _____.

    a.
    b. Image
    c.
    d.

5. _____ refers to the point in a reference frame from which measurements are made. It is the location of the zero value for each axis in the frame.

    a.
    b.
    c.
    d. Origin

ANSWER KEY
Chapter 1. SIGNALS AND SYSTEMS

1. c
2. c
3. b
4. b
5. d

You can take the complete Chapter Practice Test

for Chapter 1. SIGNALS AND SYSTEMS
on all key terms, persons, places, and concepts.

Online 99 Cents

http://www.epub115.4.2097.1.cram101.com/

Use www.Cram101.com for all your study needs

including Cram101's online interactive problem solving labs in chemistry, statistics, mathematics, and more.

CHAPTER OUTLINE: KEY TERMS, PEOPLE, PLACES, CONCEPTS
Chapter 2
LINEAR TIME-INVARIANT SYSTEMS

- System1
- Properties1
- Set1
- Core1
- Property1
- Impulse1
- Function1
- Equation1
- Negative1
- Reflection1
- Pulse1
- Amplitude1
- Stress1
- Converge1
- Weight1
- Origin1
- State1
- Axis1
- Matter1

Chapter 2. LINEAR TIME-INVARIANT SYSTEMS

Second1

Current1

Stable1

Magnitude1

Unstable1

Acceleration1

Digital1

Velocity1

Applied force1

Force1

Units1

Speed1

Constant acceleration1

Voltage1

Volt1

Capacitor1

Heat1

Time interval1

Element1

Chapter 2. LINEAR TIME-INVARIANT SYSTEMS

- Noise1
- Singularity1
- Period1
- Slope1
- Periodic1
- Complex number1
- Displacement1
- Friction1
- Charged1
- Echo1
- Sound1
- Conversion1
- Symmetry1
- Switch1
- Scalar1
- Receiver1
- Interference1
- Radar1
- Distance1

CHAPTER HIGHLIGHTS: KEY TERMS, PEOPLE, PLACES, CONCEPTS
Chapter 2. LINEAR TIME-INVARIANT SYSTEMS

System1	Defined collection of objects is called a system.
Properties1	Properties refers to qualities or attributes that, taken together, are usually unique to an object; for example, color, texture, and size.
Set1	In mathematics a set is a collection of related objects. The mathematical usage is similar to the ordinary English meaning of the word. The objects that make up a set are called the elements of the set. If a set contains an unlimited number of elements it is an infinite set. Otherwise it is a finite set.
Core1	In electromagnetism, the material inside the coils of a transformer or electromagnet is a core.
Property1	A characteristic that is inherently associated with the object which is said to have that property. For example the mass of an object is one of its properties. So also might be color, density and many other characteristics. Properties are classified as extensive or intensive. Extensive properties increase in proportion to the size of the object, as mass does for example. Intensive properties are independent of the size of the object. The density for examples remains the same if I cut an object in half and throw half of it away. Things like an object's position or velocity are not considered to be properties of the object. They are not a characteristic of the object only but are also dependent on the reference frame in which the object is located.
Impulse1	Product of force and time interval over which it acts is the impulse.
Function1	A mathematical function is a rule relating two sets of objects. Here we will restrict ourselves to objects that are numbers or vectors. One of the sets is called the domain of the function, the other is called the range of the function. Functions are frequently expressed as equations as for example Y=X+2. This function is interpreted as follows. For every X in the domain, add 2 to it to get the corresponding Y in the range. Because we are free to choose any X we want, X is called the independent variable. Because once X is chosen Y is fixed, we call Y the dependent variable.
Equation1	An equation is a mathematical expression with an equal sign in it. It signifies that the numerical or vector value on one side of the = is the same as the numerical or vector value on the other side. An equation may include variables and parameters. If any of the variables are rates of change, the equation is called a differential equation.
Negative1	The sign of the electric charge on the electron is negative.
Reflection1	Reflection refers to the change when light, sound, or other waves bounce backwards off a boundary .

Chapter 2. LINEAR TIME-INVARIANT SYSTEMS

Pulse1	A wave of short duration confined to a small portion of the medium at any given time is called a pulse. A pulse is also called a wave pulse.
Amplitude1	In any periodic motion, the maximum displacement from equilibrium is called amplitude.
Stress1	Stress is defined as force per unit area. This is one of the most basic engineering quantities.
Converge1	If the dependent variable of a function under iteration gets closer and closer to a fixed value the function is said to converge under iteration.
Weight1	Force of gravity of an object is called weight.
Origin1	Origin refers to the point in a reference frame from which measurements are made. It is the location of the zero value for each axis in the frame.
State1	Dynamical systems evolve over the course of time. The state of the system at any instant may be identified by the values of certain variables at that instant. For example specifying the angle from the vertical and the velocity of a frictionless pendulum allows us to predict its position and velocity at any future time. Therefore the state of the pendulum at any instant is its position and velocity. In this example the position and velocity are known as state variables.
Axis1	Axis refers to the imaginary line about which a planet or other object rotates .
Matter1	We call the commonly observed particles such as protons, neutrons and electrons matter particles, and their antiparticles, antimatter.
Second1	Second is a SI unit of time.
Current1	Current refers to a flow of charge. Measured in amps.
Stable1	Does not decay is stable.
Magnitude1	Magnitude refers to the size of a thing, without regard for its sign or direction. Similar to the absolute value of a number but applies to vectors as well.
Unstable1	Matter that is capable of undergoing spontaneous change, as in a radioactive nuclide or an excited nuclear system. An unstable particle is any elementary particle that spontaneously decays into other particles.

Chapter 2. LINEAR TIME-INVARIANT SYSTEMS

Acceleration1	Change in velocity divided by time interval over which it occurred is an acceleration.
Digital1	In electronics, meaning 'coded as numbers'. Digital means having discrete values as on a digital display.
Velocity1	The ratio of change in position with respect to the time interval over which the change occurred is referred to as velocity.
Applied force1	We make a somewhat arbitrary distinction among the forces acting in a dynamical system. The categories are applied force, centering force and drag or friction force. The applied force is taken to be that force which is applied to the moving parts of a system by an outside agent as for example the force applied to a pendulum by someone pushing it or the force applied to a piece of metal by the magnetic field of an electromagnet. An applied force results in an energy transfer across the system boundary.
Force1	Force refers to agent that results in accelerating or deforming an object.
9. Thus a gev is a billion electron Volts. The mass-energy of a proton or neutron is approximately 1 gev.	The units one uses should be of a size that makes sense for the particular subject at hand. It is easiest to define units in each area of science and then relate them to one another than to go around measuring particle masses in grams or cheese in proton mass units. In particle physics the standard unit is the unit of energy gev. One ev is the amount of energy that an electron gains when it moves through a potential difference of 1 Volt . G stands for Giga, or 10
Speed1	Speed refers to ratio of distance traveled to time interval.
Constant acceleration1	Acceleration that does not change in time is called constant acceleration.
Voltage1	Voltage refers to potential difference. It is a measure of the change in energy that one coulomb of electric charge undergoes when moved between 2 points.
Volt1	The unit of voltage or potential difference is a volt.
Capacitor1	Electrical device used to store charge and energy in the electrical field is referred to as a capacitor.

Chapter 2. LINEAR TIME-INVARIANT SYSTEMS

Heat1	Heat refers to quantity of energy transferred from one object to another because of a difference in temperature.
Time interval1	Difference in time between two clock readings is referred to as a time interval.
Element1	A pure substance that cannot be split up into anything simpler is called an element.
Noise1	Noise refers to sound consisting of an irregular mixture of frequencies.
Singularity1	The center of a black hole, where the curvature of spacetime is maximal is its singularity. At the singularity, the gravitational tides diverge; no solid object can even theoretically survive hitting the singularity. Although singularities generally predict inconsistencies in theory, singularities within black holes do not necessarily imply that general relativity is incomplete so long as singularities are always surrounded by event horizons. A proper formulation of quantum gravity may well avoid the classical singularity at the centers of black holes.
Period1	Period refers to time needed to repeat one complete cycle of motion.
Slope1	Ratio of the vertical separation, or rise to the horizontal separation, or run are called the slope.
Periodic1	A dynamical system or function which at some point returns to the same state or value. If a system or function ever revisits the identical state or value it will continue to come back to it again and again in equal intervals of time. That is why we call such a system periodic.
Complex number1	Ordinary numbers that we use for accounting and simple calculations are called real numbers. There is no real number such that the square of that number is -1 since the product of any real number with itself is positive. To remedy this situation, the square root of -1 was defined and given the symbol i. Then a new class of numbers was invented, called imaginary numbers, made up of i times the set of real numbers. A complex number is a number comprised of a real part added to an imaginary part like a+b*i.
Displacement1	Displacement refers to change in position. A vector quantity.
Friction1	Force opposing relative motion of two objects are in contact is friction.
Charged1	Object that has an unbalance of positive and negative electrical charges is referred to as charged.
Echo1	A reflection of a sound or ultrasound is called echo.

Chapter 2. LINEAR TIME-INVARIANT SYSTEMS

Sound1	A longitudinal wave, usually through the air but can travel through liquids and solids is called sound.
Conversion1	Chemical process turning U308 into UF6 preparatory to enrichment is conversion.
Symmetry1	Symmetry refers to property that is now charged when operation or reference frame is charged.
Switch1	Switch refers to a device in an electrical circuit that breaks or makes a complete path for the current.
Scalar1	Quantity, like distance, that has only a magnitude, or size is referred to as scalar.
Receiver1	Device that detects electromagnetic waves is called receiver.
Interference1	Phenomenon of light where the relative phase difference between two light waves produces light or dark spots, a result of light's wavelike nature is called the interference.
Radar1	A method of detecting, locating, or tracking an object by using beamed, reflected, and timed radio waves. RADAR also refers to the electronic equipment that uses radio waves to detect, locate, and track objects.
Distance1	Distance refers to separation between two points. A scalar quantity.

PRACTICE QUIZ
Chapter 2. LINEAR TIME-INVARIANT SYSTEMS

1. _____ refers to the size of a thing, without regard for its sign or direction. Similar to the absolute value of a number but applies to vectors as well.

    a.
    b.
    c.
    d. Magnitude

2. Dynamical systems evolve over the course of time. The _____ of the system at any instant may be identified by the values of certain variables at that instant. For example specifying the angle from the vertical and the velocity of a frictionless pendulum allows us to predict its position and velocity at any future time. Therefore the _____ of the pendulum at any instant is its position and velocity. In this example the position and velocity are known as _____ variables.

    a.
    b. State
    c.
    d.

3. We make a somewhat arbitrary distinction among the forces acting in a dynamical system. The categories are applied force, centering force and drag or friction force. The _____ is taken to be that force which is applied to the moving parts of a system by an outside agent as for example the force applied to a pendulum by someone pushing it or the force applied to a piece of metal by the magnetic field of an electromagnet. An _____ results in an energy transfer across the system boundary.

    a. Applied force
    b.
    c.
    d.

4. Matter that is capable of undergoing spontaneous change, as in a radioactive nuclide or an excited nuclear system. An _____ particle is any elementary particle that spontaneously decays into other particles.

    a.
    b.
    c.
    d. Unstable

----

5. In mathematics a _____ is a collection of related objects. The mathematical usage is similar to the ordinary English meaning of the word. The objects that make up a _____ are called the elements of the set. If a _____ contains an unlimited number of elements it is an infinite set. Otherwise it is a finite set.

   a.
   b.
   c.
   d. Set

ANSWER KEY
Chapter 2. LINEAR TIME-INVARIANT SYSTEMS

1. d
2. b
3. a
4. d
5. d

You can take the complete Chapter Practice Test

for Chapter 2. LINEAR TIME-INVARIANT SYSTEMS
on all key terms, persons, places, and concepts.

Online 99 Cents

http://www.epub115.4.2097.2.cram101.com/

Use www.Cram101.com for all your study needs

including Cram101's online interactive problem solving labs in chemistry, statistics, mathematics, and more.

CHAPTER OUTLINE: KEY TERMS, PEOPLE, PLACES, CONCEPTS
Chapter 3
FOURIER SERIES REPRESENTATION OF PERIODIC SIGNALS

- Periodic1
- Set1
- Property1
- System1
- Impulse1
- Current1
- Properties1
- Energy1
- Distance1
- Normal1
- Position1
- Solid1
- Second1
- Digital1
- Magnitude1
- Amplitude1
- Function1
- Complex number1
- Range1

Chapter 3. FOURIER SERIES REPRESENTATION OF PERIODIC SIGNALS

- Equation1
- Frequency1
- Domain1
- Period1
- Fundamental frequency1
- First harmonic1
- Light1
- Intensity1
- Phase1
- Wave1
- Symmetry1
- Diverge1
- Converge1
- Boundary1
- Matter1
- Proof1
- Negative1
- Power1
- Stable1

Chapter 3. FOURIER SERIES REPRESENTATION OF PERIODIC SIGNALS

- Unstable1
- Units1
- Axis1
- Switch1
- Image1
- Decibel1
- Harmonics1
- Gradient1
- Noise1
- Modulation1
- Receiver1
- Voltage1
- Capacitor1
- Resistor1
- Attenuation1
- Speed1
- Pulse1
- Periodic wave1
- Time interval1

Chapter 3. FOURIER SERIES REPRESENTATION OF PERIODIC SIGNALS

Physics1

Heat1

Diffusion1

Temperature1

Vector1

Velocity1

Force1

CHAPTER HIGHLIGHTS: KEY TERMS, PEOPLE, PLACES, CONCEPTS
Chapter 3. FOURIER SERIES REPRESENTATION OF PERIODIC SIGNALS

Periodic1	A dynamical system or function which at some point returns to the same state or value. If a system or function ever revisits the identical state or value it will continue to come back to it again and again in equal intervals of time. That is why we call such a system periodic.
Set1	In mathematics a set is a collection of related objects. The mathematical usage is similar to the ordinary English meaning of the word. The objects that make up a set are called the elements of the set. If a set contains an unlimited number of elements it is an infinite set. Otherwise it is a finite set.
Property1	A characteristic that is inherently associated with the object which is said to have that property. For example the mass of an object is one of its properties. So also might be color, density and many other characteristics. Properties are classified as extensive or intensive. Extensive properties increase in proportion to the size of the object, as mass does for example. Intensive properties are independent of the size of the object. The density for examples remains the same if I cut an object in half and throw half of it away. Things like an object's position or velocity are not considered to be properties of the object. They are not a characteristic of the object only but are also dependent on the reference frame in which the object is located.
System1	Defined collection of objects is called a system.
Impulse1	Product of force and time interval over which it acts is the impulse.
Current1	Current refers to a flow of charge. Measured in amps.
Properties1	Properties refers to qualities or attributes that, taken together, are usually unique to an object; for example, color, texture, and size.
Energy1	Energy refers to non-material property capable of causing changes in matter.
Distance1	Distance refers to separation between two points. A scalar quantity.
Normal1	Perpendicular to plane of interest is called normal.
Position1	Separation between object and a reference point is a position.
Solid1	State of matter with fixed volume and shape is referred to as solid.
Second1	Second is a SI unit of time.

Chapter 3. FOURIER SERIES REPRESENTATION OF PERIODIC SIGNALS

Digital1	In electronics, meaning 'coded as numbers'. Digital means having discrete values as on a digital display.
Magnitude1	Magnitude refers to the size of a thing, without regard for its sign or direction. Similar to the absolute value of a number but applies to vectors as well.
Amplitude1	In any periodic motion, the maximum displacement from equilibrium is called amplitude.
Function1	A mathematical function is a rule relating two sets of objects. Here we will restrict ourselves to objects that are numbers or vectors. One of the sets is called the domain of the function, the other is called the range of the function. Functions are frequently expressed as equations as for example Y=X+2. This function is interpreted as follows. For every X in the domain, add 2 to it to get the corresponding Y in the range. Because we are free to choose any X we want, X is called the independent variable. Because once X is chosen Y is fixed, we call Y the dependent variable.
Complex number1	Ordinary numbers that we use for accounting and simple calculations are called real numbers. There is no real number such that the square of that number is -1 since the product of any real number with itself is positive. To remedy this situation, the square root of -1 was defined and given the symbol i. Then a new class of numbers was invented, called imaginary numbers, made up of i times the set of real numbers. A complex number is a number comprised of a real part added to an imaginary part like a+b*i.
Range1	The range is the set of values that the dependent variable of a function may take on. A range may be finite as in the set of numbers {1, 2, 3.n} or infinite as in all the mumbers between 0 and 1.
Equation1	An equation is a mathematical expression with an equal sign in it. It signifies that the numerical or vector value on one side of the = is the same as the numerical or vector value on the other side. An equation may include variables and parameters. If any of the variables are rates of change, the equation is called a differential equation.
Frequency1	Frequency refers to number of occurrences per unit time.
Domain1	The domain is the set of values that the independent variable of a function may take on. A domain may be finite as in the set of numbers {1, 2, 3.n} or infinite as in all the numbers between 0 and 1.
Period1	Period refers to time needed to repeat one complete cycle of motion.

Chapter 3. FOURIER SERIES REPRESENTATION OF PERIODIC SIGNALS

Fundamental frequency1	Fundamental frequency refers to the lowest frequency that can set up standing waves in an air column or on a string.
First harmonic1	First harmonic in music, the fundamental frequency.
Light1	Electromagnetic radiation with wavelengths between 400 and 700 nm that is visible is called light.
Intensity1	The amount of radiation, for example, the number of photons arriving in a given time, is called intensity.
Phase1	The particles in a wave, which are in the same state of vibration, i.e., the same position and the same direction of motion are said to be in the same phase.
Wave1	A set of oscillations or vibrations that transfer energy without any transfer of mass is the wave.
Symmetry1	Symmetry refers to property that is now charged when operation or reference frame is charged.
Diverge1	If the dependent variable of a function under iteration increases without limit the function is said to diverge under iteration.
Converge1	If the dependent variable of a function under iteration gets closer and closer to a fixed value the function is said to converge under iteration.
Boundary1	Boundary refers to the division between two regions of differing physical properties.
Matter1	We call the commonly observed particles such as protons, neutrons and electrons matter particles, and their antiparticles, antimatter.
Proof1	A measure of ethanol concentration of an alcoholic beverage; proof is double the concentration by volume; for example, 50 percent by volume is 100 proof.
Negative1	The sign of the electric charge on the electron is negative.
Power1	Power refers to rate of doing work; rate of energy conversion.
Stable1	Does not decay is stable.

Chapter 3. FOURIER SERIES REPRESENTATION OF PERIODIC SIGNALS

Unstable1	Matter that is capable of undergoing spontaneous change, as in a radioactive nuclide or an excited nuclear system. An unstable particle is any elementary particle that spontaneously decays into other particles.
9. Thus a gev is a billion electron Volts. The mass-energy of a proton or neutron is approximately 1 gev.	The units one uses should be of a size that makes sense for the particular subject at hand. It is easiest to define units in each area of science and then relate them to one another than to go around measuring particle masses in grams or cheese in proton mass units. In particle physics the standard unit is the unit of energy gev. One ev is the amount of energy that an electron gains when it moves through a potential difference of 1 Volt . G stands for Giga, or 10
Axis1	Axis refers to the imaginary line about which a planet or other object rotates .
Switch1	Switch refers to a device in an electrical circuit that breaks or makes a complete path for the current.
Image1	Reproduction of object formed with lenses or mirrors is the image.
Decibel1	Unit of sound level is a decibel.
Harmonics1	Frequencies produced by musical instrument that are multiples of fundamental tone are referred to as harmonics.
Gradient1	Gradient refers to the slope of a graph.
Noise1	Noise refers to sound consisting of an irregular mixture of frequencies.
Modulation1	The process of superimposing information onto a carrier wave is referred to as modulation.
Receiver1	Device that detects electromagnetic waves is called receiver.
Voltage1	Voltage refers to potential difference. It is a measure of the change in energy that one coulomb of electric charge undergoes when moved between 2 points.
Capacitor1	Electrical device used to store charge and energy in the electrical field is referred to as a capacitor.
Resistor1	Device designed to have a specific resistance is a resistor.

Chapter 3. FOURIER SERIES REPRESENTATION OF PERIODIC SIGNALS

Attenuation1	The process by which a compound is reduced in concentration over time, through adsorption, degradation, dilution, and/or transformation. Radiologically, it is the reduction of the intensity of radiation upon passage through a medium. The attenuation is caused by absorption and scattering.
Speed1	Speed refers to ratio of distance traveled to time interval.
Pulse1	A wave of short duration confined to a small portion of the medium at any given time is called a pulse. A pulse is also called a wave pulse.
Periodic wave1	A wave in which the particles of the medium oscillate continuously about their mean positions regularly at fixed intervals of time is called a periodic wave.
Time interval1	Difference in time between two clock readings is referred to as a time interval.
Physics1	Study of matter and energy and their relationship is physics.
Heat1	Heat refers to quantity of energy transferred from one object to another because of a difference in temperature.
Diffusion1	The spreading out of a substance, due to the kinetic energy of its particles, to fill all of the available space are called diffusion.
Temperature1	Temperature refers to measure of hotness of object on a quantitative scale. In gases, proportional to average kinetic energy of molecules.
Vector1	Any quantity that has both magnitude and direction. Velocity is a vector.
Velocity1	The ratio of change in position with respect to the time interval over which the change occurred is referred to as velocity.
Force1	Force refers to agent that results in accelerating or deforming an object.

PRACTICE QUIZ
Chapter 3. FOURIER SERIES REPRESENTATION OF PERIODIC SIGNALS

1. If the dependent variable of a function under iteration gets closer and closer to a fixed value the function is said to _____ under iteration.

    a.
    b. Converge
    c.
    d.

2. Perpendicular to plane of interest is called _____.

    a.
    b.
    c.
    d. Normal

3. A dynamical system or function which at some point returns to the same state or value. If a system or function ever revisits the identical state or value it will continue to come back to it again and again in equal intervals of time. That is why we call such a system _____.

    a. Periodic
    b.
    c.
    d.

4. _____ refers to property that is now charged when operation or reference frame is charged.

    a.
    b.
    c. Symmetry
    d.

ANSWER KEY
Chapter 3. FOURIER SERIES REPRESENTATION OF PERIODIC SIGNALS

1. b
2. d
3. a
4. c

You can take the complete Chapter Practice Test

for Chapter 3. FOURIER SERIES REPRESENTATION OF PERIODIC SIGNALS
on all key terms, persons, places, and concepts.

Online 99 Cents

http://www.epub115.4.2097.3.cram101.com/

Use www.Cram101.com for all your study needs

including Cram101's online interactive problem solving labs in chemistry, statistics, mathematics, and more.

CHAPTER OUTLINE: KEY TERMS, PEOPLE, PLACES, CONCEPTS
Chapter 4
THE CONTINUOUS - TIME FOURIER TRANSFORM

- Periodic 1
- Energy 1
- Frequency 1
- Period 1
- Fundamental frequency 1
- Properties 1
- Wave 1
- Function 1
- Pulse 1
- Domain 1
- Set 1
- Equation 1
- Range 1
- Impulse 1
- Magnitude 1
- Phase 1
- Amplitude 1
- Property 1
- Inverse relationship 1

Chapter 4. THE CONTINUOUS - TIME FOURIER TRANSFORM

- Modulation1
- Proof1
- Symmetry1
- Negative1
- Image1
- Speed1
- Compression1
- Sound1
- Uncertainty principle1
- Physics1
- Power1
- Second1
- System1
- Attenuation1
- Stable1
- Unstable1
- Element1
- Sine1
- Reflection1

Chapter 4. THE CONTINUOUS - TIME FOURIER TRANSFORM

Slope1

Current1

Echo1

Temperature1

Liquid1

Mercury1

Thermometer1

Interference1

Noise1

CHAPTER HIGHLIGHTS: KEY TERMS, PEOPLE, PLACES, CONCEPTS
Chapter 4. THE CONTINUOUS - TIME FOURIER TRANSFORM

Periodic1	A dynamical system or function which at some point returns to the same state or value. If a system or function ever revisits the identical state or value it will continue to come back to it again and again in equal intervals of time. That is why we call such a system periodic.
Energy1	Energy refers to non-material property capable of causing changes in matter.
Frequency1	Frequency refers to number of occurrences per unit time.
Period1	Period refers to time needed to repeat one complete cycle of motion.
Fundamental frequency1	Fundamental frequency refers to the lowest frequency that can set up standing waves in an air column or on a string .
Properties1	Properties refers to qualities or attributes that, taken together, are usually unique to an object; for example, color, texture, and size.
Wave1	A set of oscillations or vibrations that transfer energy without any transfer of mass is the wave.
Function1	A mathematical function is a rule relating two sets of objects. Here we will restrict ourselves to objects that are numbers or vectors. One of the sets is called the domain of the function, the other is called the range of the function. Functions are frequently expressed as equations as for example Y=X+2. This function is interpreted as follows. For every X in the domain, add 2 to it to get the corresponding Y in the range. Because we are free to choose any X we want, X is called the independent variable. Because once X is chosen Y is fixed, we call Y the dependent variable.
Pulse1	A wave of short duration confined to a small portion of the medium at any given time is called a pulse. A pulse is also called a wave pulse.
Domain1	The domain is the set of values that the independent variable of a function may take on. A domain may be finite as in the set of numbers {1, 2, 3.n} or infinite as in all the numbers between 0 and 1.
Set1	In mathematics a set is a collection of related objects. The mathematical usage is similar to the ordinary English meaning of the word. The objects that make up a set are called the elements of the set. If a set contains an unlimited number of elements it is an infinite set. Otherwise it is a finite set.

Chapter 4. THE CONTINUOUS - TIME FOURIER TRANSFORM

Equation1	An equation is a mathematical expression with an equal sign in it. It signifies that the numerical or vector value on one side of the = is the same as the numerical or vector value on the other side. An equation may include variables and parameters. If any of the variables are rates of change, the equation is called a differential equation.
Range1	The range is the set of values that the dependent variable of a function may take on. A range may be finite as in the set of numbers {1, 2, 3.n} or infinite as in all the mumbers between 0 and 1.
Impulse1	Product of force and time interval over which it acts is the impulse.
Magnitude1	Magnitude refers to the size of a thing, without regard for its sign or direction. Similar to the absolute value of a number but applies to vectors as well.
Phase1	The particles in a wave, which are in the same state of vibration, i.e., the same position and the same direction of motion are said to be in the same phase.
Amplitude1	In any periodic motion, the maximum displacement from equilibrium is called amplitude.
Property1	A characteristic that is inherently associated with the object which is said to have that property. For example the mass of an object is one of its properties. So also might be color, density and many other characteristics. Properties are classified as extensive or intensive. Extensive properties increase in proportion to the size of the object, as mass does for example. Intensive properties are independent of the size of the object. The density for examples remains the same if I cut an object in half and throw half of it away. Things like an object's position or velocity are not considered to be properties of the object. They are not a characteristic of the object only but are also dependent on the reference frame in which the object is located.
Inverse relationship1	Inverse relationship refers to mathematical relationship between two variables, x and y, summarized by the equation xy=k, where k is a constant.
Modulation1	The process of superimposing information onto a carrier wave is referred to as modulation.
Proof1	A measure of ethanol concentration of an alcoholic beverage; proof is double the concentration by volume; for example, 50 percent by volume is 100 proof.
Symmetry1	Symmetry refers to property that is now charged when operation or reference frame is charged.

Chapter 4. THE CONTINUOUS - TIME FOURIER TRANSFORM

Negative1	The sign of the electric charge on the electron is negative.
Image1	Reproduction of object formed with lenses or mirrors is the image.
Speed1	Speed refers to ratio of distance traveled to time interval.
Compression1	In a sound wave, a compression is a region in the material that transmits the wave where the particles are closer together than normal.
Sound1	A longitudinal wave, usually through the air but can travel through liquids and solids is called sound.
Uncertainty principle1	Uncertainty principle refers to a principle, central to quantum mechanics, which states that two complementary parameters cannot both be known to infinite accuracy; the more you know about one, the less you know about the other.
Physics1	Study of matter and energy and their relationship is physics.
Power1	Power refers to rate of doing work; rate of energy conversion.
Second1	Second is a SI unit of time.
System1	Defined collection of objects is called a system.
Attenuation1	The process by which a compound is reduced in concentration over time, through adsorption, degradation, dilution, and/or transformation. Radiologically, it is the reduction of the intensity of radiation upon passage through a medium. The attenuation is caused by absorption and scattering.
Stable1	Does not decay is stable.
Unstable1	Matter that is capable of undergoing spontaneous change, as in a radioactive nuclide or an excited nuclear system. An unstable particle is any elementary particle that spontaneously decays into other particles.
Element1	A pure substance that cannot be split up into anything simpler is called an element.
Sine1	Sine refers to the ratio of the opposite side and the hypotenuse.

Chapter 4. THE CONTINUOUS - TIME FOURIER TRANSFORM

Reflection1	Reflection refers to the change when light, sound, or other waves bounce backwards off a boundary .
Slope1	Ratio of the vertical separation, or rise to the horizontal separation, or run are called the slope.
Current1	Current refers to a flow of charge. Measured in amps.
Echo1	A reflection of a sound or ultrasound is called echo.
Temperature1	Temperature refers to measure of hotness of object on a quantitative scale. In gases, proportional to average kinetic energy of molecules.
Liquid1	Material that has fixed volume but whose shape depends on the container is called liquid.
Mercury1	Mercury refers to the innermost planet in the solar system, and a metallic element that is liquid at room temperature.
Thermometer1	Device used to measure temperature is a thermometer.
Interference1	Phenomenon of light where the relative phase difference between two light waves produces light or dark spots, a result of light's wavelike nature is called the interference.
Noise1	Noise refers to sound consisting of an irregular mixture of frequencies.

PRACTICE QUIZ
Chapter 4. THE CONTINUOUS - TIME FOURIER TRANSFORM

1. In a sound wave, a _____ is a region in the material that transmits the wave where the particles are closer together than normal.

    a. Compression
    b.
    c.
    d.

2. In mathematics a _____ is a collection of related objects. The mathematical usage is similar to the ordinary English meaning of the word. The objects that make up a _____ are called the elements of the set. If a _____ contains an unlimited number of elements it is an infinite set. Otherwise it is a finite set.

    a. Set
    b.
    c.
    d.

3. A dynamical system or function which at some point returns to the same state or value. If a system or function ever revisits the identical state or value it will continue to come back to it again and again in equal intervals of time. That is why we call such a system _____.

    a. Periodic
    b.
    c.
    d.

4. _____ refers to non-material property capable of causing changes in matter.

    a.
    b.
    c.
    d. Energy

5. _____ refers to number of occurrences per unit time.

    a.
    b.
    c.
    d. Frequency

ANSWER KEY
Chapter 4. THE CONTINUOUS - TIME FOURIER TRANSFORM

1. a
2. a
3. a
4. d
5. d

You can take the complete Chapter Practice Test

for Chapter 4. THE CONTINUOUS - TIME FOURIER TRANSFORM
on all key terms, persons, places, and concepts.

Online 99 Cents

http://www.epub115.4.2097.4.cram101.com/

Use www.Cram101.com for all your study needs

including Cram101's online interactive problem solving labs in chemistry, statistics, mathematics, and more.

CHAPTER OUTLINE: KEY TERMS, PEOPLE, PLACES, CONCEPTS
Chapter 5
THE DISCRETE-TIME FOURIER TRANSFORM

- Properties1
- Current1
- Periodic1
- Wave1
- Function1
- Period1
- Property1
- Range1
- Second1
- Frequency1
- Domain1
- Equation1
- Magnitude1
- Phase1
- Pulse1
- Impulse1
- Fundamental frequency1
- State1
- System1

Chapter 5. THE DISCRETE-TIME FOURIER TRANSFORM

- Symmetry 1
- Phase angle 1
- Speed 1
- Inverse relationship 1
- Quantity 1
- Energy 1
- Power 1
- Amplitude 1
- Stable 1
- Converge 1
- Mechanics 1
- Sine 1
- Set 1
- Gravity 1
- Digital 1
- Time interval 1
- Cosine 1
- Independent variable 1

CHAPTER HIGHLIGHTS: KEY TERMS, PEOPLE, PLACES, CONCEPTS
Chapter 5. THE DISCRETE-TIME FOURIER TRANSFORM

Properties1	Properties refers to qualities or attributes that, taken together, are usually unique to an object; for example, color, texture, and size.
Current1	Current refers to a flow of charge. Measured in amps.
Periodic1	A dynamical system or function which at some point returns to the same state or value. If a system or function ever revisits the identical state or value it will continue to come back to it again and again in equal intervals of time. That is why we call such a system periodic.
Wave1	A set of oscillations or vibrations that transfer energy without any transfer of mass is the wave.
Function1	A mathematical function is a rule relating two sets of objects. Here we will restrict ourselves to objects that are numbers or vectors. One of the sets is called the domain of the function, the other is called the range of the function. Functions are frequently expressed as equations as for example Y=X+2. This function is interpreted as follows. For every X in the domain, add 2 to it to get the corresponding Y in the range. Because we are free to choose any X we want, X is called the independent variable. Because once X is chosen Y is fixed, we call Y the dependent variable.
Period1	Period refers to time needed to repeat one complete cycle of motion.
Property1	A characteristic that is inherently associated with the object which is said to have that property. For example the mass of an object is one of its properties. So also might be color, density and many other characteristics. Properties are classified as extensive or intensive. Extensive properties increase in proportion to the size of the object, as mass does for example. Intensive properties are independent of the size of the object. The density for examples remains the same if I cut an object in half and throw half of it away. Things like an object's position or velocity are not considered to be properties of the object. They are not a characteristic of the object only but are also dependent on the reference frame in which the object is located.
Range1	The range is the set of values that the dependent variable of a function may take on. A range may be finite as in the set of numbers {1, 2, 3.n} or infinite as in all the mumbers between 0 and 1.
Second1	Second is a SI unit of time.
Frequency1	Frequency refers to number of occurrences per unit time.

Chapter 5. THE DISCRETE-TIME FOURIER TRANSFORM

Domain1	The domain is the set of values that the independent variable of a function may take on. A domain may be finite as in the set of numbers {1, 2, 3.n} or infinite as in all the numbers between 0 and 1.
Equation1	An equation is a mathematical expression with an equal sign in it. It signifies that the numerical or vector value on one side of the = is the same as the numerical or vector value on the other side. An equation may include variables and parameters. If any of the variables are rates of change, the equation is called a differential equation.
Magnitude1	Magnitude refers to the size of a thing, without regard for its sign or direction. Similar to the absolute value of a number but applies to vectors as well.
Phase1	The particles in a wave, which are in the same state of vibration, i.e., the same position and the same direction of motion are said to be in the same phase.
Pulse1	A wave of short duration confined to a small portion of the medium at any given time is called a pulse. A pulse is also called a wave pulse.
Impulse1	Product of force and time interval over which it acts is the impulse.
Fundamental frequency1	Fundamental frequency refers to the lowest frequency that can set up standing waves in an air column or on a string .
State1	Dynamical systems evolve over the course of time. The state of the system at any instant may be identified by the values of certain variables at that instant. For example specifying the angle from the vertical and the velocity of a frictionless pendulum allows us to predict its position and velocity at any future time. Therefore the state of the pendulum at any instant is its position and velocity. In this example the position and velocity are known as state variables.
System1	Defined collection of objects is called a system.
Symmetry1	Symmetry refers to property that is now charged when operation or reference frame is charged.
Phase angle1	The phase angle is the offset from the origin of a periodic function like the sine or cosine.
Speed1	Speed refers to ratio of distance traveled to time interval.

Chapter 5. THE DISCRETE-TIME FOURIER TRANSFORM

Inverse relationship1	Inverse relationship refers to mathematical relationship between two variables, x and y, summarized by the equation xy=k, where k is a constant.
Quantity1	A numerical value either scalar or vector, which describes some attribute of an object like its position or its velocity . We sometimes speak of physical quantities to signify that we are talking about an object's properties or attributes as opposed to a purely mathematical quantity.
Energy1	Energy refers to non-material property capable of causing changes in matter.
Power1	Power refers to rate of doing work; rate of energy conversion.
Amplitude1	In any periodic motion, the maximum displacement from equilibrium is called amplitude.
Stable1	Does not decay is stable.
Converge1	If the dependent variable of a function under iteration gets closer and closer to a fixed value the function is said to converge under iteration.
Mechanics1	The study of objects in motion. Mechanics is normally limited to a small number of large slow objects, as opposed to statistical mechanics which deals with large numbers of objects, relativistic mechanics which deals with objects moving near the speed of light and quantum mechanics which deals with objects more or less the size of atoms. Mechanics encompasses the topics of kinematics and dynamics .
Sine1	Sine refers to the ratio of the opposite side and the hypotenuse.
Set1	In mathematics a set is a collection of related objects. The mathematical usage is similar to the ordinary English meaning of the word. The objects that make up a set are called the elements of the set. If a set contains an unlimited number of elements it is an infinite set. Otherwise it is a finite set.
Gravity1	A force with infinite range which acts between objects, such as planets,according to their mass is called gravity.
Digital1	In electronics, meaning 'coded as numbers'. Digital means having discrete values as on a digital display.
Time interval1	Difference in time between two clock readings is referred to as a time interval.

Chapter 5. THE DISCRETE-TIME FOURIER TRANSFORM

Cosine1	The ratio of the adjacent side to the hypotenuse is the cosine.
Independent variable1	Variable that is manipulated or changed in an experiment is referred to as the independent variable.

PRACTICE QUIZ
Chapter 5. THE DISCRETE-TIME FOURIER TRANSFORM

1. Defined collection of objects is called a _____.

   a.
   b.
   c. System
   d.

2. _____ refers to property that is now charged when operation or reference frame is charged.

   a.
   b.
   c. Symmetry
   d.

3. A mathematical _____ is a rule relating two sets of objects. Here we will restrict ourselves to objects that are numbers or vectors. One of the sets is called the domain of the function, the other is called the range of the function. Functions are frequently expressed as equations as for example Y=X+2. This _____ is interpreted as follows. For every X in the domain, add 2 to it to get the corresponding Y in the range. Because we are free to choose any X we want, X is called the independent variable. Because once X is chosen Y is fixed, we call Y the dependent variable.

   a.
   b. Function
   c.
   d.

4. _____ refers to qualities or attributes that, taken together, are usually unique to an object; for example, color, texture, and size.

   a. Properties
   b.
   c.
   d.

5. _____ refers to a flow of charge. Measured in amps.

   a.
   b.
   c. Current
   d.

ANSWER KEY
Chapter 5. THE DISCRETE-TIME FOURIER TRANSFORM

1. c
2. c
3. b
4. a
5. c

You can take the complete Chapter Practice Test

for Chapter 5. THE DISCRETE-TIME FOURIER TRANSFORM
on all key terms, persons, places, and concepts.

Online 99 Cents

http://www.epub115.4.2097.5.cram101.com/

Use www.Cram101.com for all your study needs

including Cram101's online interactive problem solving labs in chemistry, statistics, mathematics, and more.

CHAPTER OUTLINE: KEY TERMS, PEOPLE, PLACES, CONCEPTS
Chapter 6
TIME AND FREQUENCY CHARACTERIZATION OF SIGNALS AND SYSTEMS

- Frequency1
- System1
- Domain1
- Impulse1
- Magnitude1
- Phase1
- Equation1
- Energy1
- Phase angle1
- Function1
- Wave1
- Amplitude1
- Property1
- Sound1
- Rad1
- Negative1
- Phase change1
- Second1
- Image1

Chapter 6. TIME AND FREQUENCY CHARACTERIZATION OF SIGNALS AND SYSTEMS

- Intensity1
- Axis1
- Set1
- Slope1
- Hertz1
- Dispersion1
- Current1
- Units1
- Range1
- Attenuation1
- Origin1
- Power1
- Decibel1
- Symmetry1
- Properties1
- Periodic1
- Inverse relationship1
- Time interval1
- Quantity1

Chapter 6. TIME AND FREQUENCY CHARACTERIZATION OF SIGNALS AND SYSTEMS

- Core1
- Speed1
- Applied force1
- Force1
- Displacement1
- Equilibrium1
- Position1
- Restoring force1
- Natural frequency1
- Decay1
- Oscillation1
- Solid1
- Unstable1
- Stable1
- Converge1
- Absorber1
- Distance1
- Excitation1
- Pulse1

Chapter 6. TIME AND FREQUENCY CHARACTERIZATION OF SIGNALS AND SYSTEMS

- Period1
- Digital1
- Voltage1
- Capacitor1
- Sine1
- Electric motor1
- Angular velocity1
- Velocity1
- State1
- Modulation1
- Noise1
- Friction1
- Element1

CHAPTER HIGHLIGHTS: KEY TERMS, PEOPLE, PLACES, CONCEPTS
Chapter 6. TIME AND FREQUENCY CHARACTERIZATION OF SIGNALS AND SYSTEMS

Frequency1	Frequency refers to number of occurrences per unit time.
System1	Defined collection of objects is called a system.
Domain1	The domain is the set of values that the independent variable of a function may take on. A domain may be finite as in the set of numbers {1, 2, 3.n} or infinite as in all the numbers between 0 and 1.
Impulse1	Product of force and time interval over which it acts is the impulse.
Magnitude1	Magnitude refers to the size of a thing, without regard for its sign or direction. Similar to the absolute value of a number but applies to vectors as well.
Phase1	The particles in a wave, which are in the same state of vibration, i.e., the same position and the same direction of motion are said to be in the same phase.
Equation1	An equation is a mathematical expression with an equal sign in it. It signifies that the numerical or vector value on one side of the = is the same as the numerical or vector value on the other side. An equation may include variables and parameters. If any of the variables are rates of change, the equation is called a differential equation.
Energy1	Energy refers to non-material property capable of causing changes in matter.
Phase angle1	The phase angle is the offset from the origin of a periodic function like the sine or cosine.
Function1	A mathematical function is a rule relating two sets of objects. Here we will restrict ourselves to objects that are numbers or vectors. One of the sets is called the domain of the function, the other is called the range of the function. Functions are frequently expressed as equations as for example Y=X+2. This function is interpreted as follows. For every X in the domain, add 2 to it to get the corresponding Y in the range. Because we are free to choose any X we want, X is called the independent variable. Because once X is chosen Y is fixed, we call Y the dependent variable.
Wave1	A set of oscillations or vibrations that transfer energy without any transfer of mass is the wave.
Amplitude1	In any periodic motion, the maximum displacement from equilibrium is called amplitude.

Chapter 6. TIME AND FREQUENCY CHARACTERIZATION OF SIGNALS AND SYSTEMS

Property1	A characteristic that is inherently associated with the object which is said to have that property. For example the mass of an object is one of its properties. So also might be color, density and many other characteristics. Properties are classified as extensive or intensive. Extensive properties increase in proportion to the size of the object, as mass does for example. Intensive properties are independent of the size of the object. The density for examples remains the same if I cut an object in half and throw half of it away. Things like an object's position or velocity are not considered to be properties of the object. They are not a characteristic of the object only but are also dependent on the reference frame in which the object is located.
Sound1	A longitudinal wave, usually through the air but can travel through liquids and solids is called sound.
Rad1	One rad is equal to an energy absorption of 100 ergs in a gram of any material. An 'erg' is a unit for quantifying energy .
Negative1	The sign of the electric charge on the electron is negative.
Phase change1	The action of a substance changing from one state of matter to another; a phase change always absorbs or releases internal potential energy that is not associated with a temperature change .
Second1	Second is a SI unit of time.
Image1	Reproduction of object formed with lenses or mirrors is the image.
Intensity1	The amount of radiation, for example, the number of photons arriving in a given time, is called intensity.
Axis1	Axis refers to the imaginary line about which a planet or other object rotates .
Set1	In mathematics a set is a collection of related objects. The mathematical usage is similar to the ordinary English meaning of the word. The objects that make up a set are called the elements of the set. If a set contains an unlimited number of elements it is an infinite set. Otherwise it is a finite set.
Slope1	Ratio of the vertical separation, or rise to the horizontal separation, or run are called the slope.
Hertz1	Unit of frequency equal to one event or cycle per second is called hertz.

Chapter 6. TIME AND FREQUENCY CHARACTERIZATION OF SIGNALS AND SYSTEMS

Dispersion1	The splitting of light into its constituent colours is called dispersion.
Current1	Current refers to a flow of charge. Measured in amps.
9. Thus a gev is a billion electron Volts. The mass-energy of a proton or neutron is approximately 1 gev.	The units one uses should be of a size that makes sense for the particular subject at hand. It is easiest to define units in each area of science and then relate them to one another than to go around measuring particle masses in grams or cheese in proton mass units. In particle physics the standard unit is the unit of energy gev. One ev is the amount of energy that an electron gains when it moves through a potential difference of 1 Volt . G stands for Giga, or 10
Range1	The range is the set of values that the dependent variable of a function may take on. A range may be finite as in the set of numbers {1, 2, 3.n} or infinite as in all the mumbers between 0 and 1.
Attenuation1	The process by which a compound is reduced in concentration over time, through adsorption, degradation, dilution, and/or transformation. Radiologically, it is the reduction of the intensity of radiation upon passage through a medium. The attenuation is caused by absorption and scattering.
Origin1	Origin refers to the point in a reference frame from which measurements are made. It is the location of the zero value for each axis in the frame.
Power1	Power refers to rate of doing work; rate of energy conversion.
Decibel1	Unit of sound level is a decibel.
Symmetry1	Symmetry refers to property that is now charged when operation or reference frame is charged.
Properties1	Properties refers to qualities or attributes that, taken together, are usually unique to an object; for example, color, texture, and size.
Periodic1	A dynamical system or function which at some point returns to the same state or value. If a system or function ever revisits the identical state or value it will continue to come back to it again and again in equal intervals of time. That is why we call such a system periodic.

Chapter 6. TIME AND FREQUENCY CHARACTERIZATION OF SIGNALS AND SYSTEMS

Inverse relationship1	Inverse relationship refers to mathematical relationship between two variables, x and y, summarized by the equation xy=k, where k is a constant.
Time interval1	Difference in time between two clock readings is referred to as a time interval.
Quantity1	A numerical value either scalar or vector, which describes some attribute of an object like its position or its velocity . We sometimes speak of physical quantities to signify that we are talking about an object's properties or attributes as opposed to a purely mathematical quantity.
Core1	In electromagnetism, the material inside the coils of a transformer or electromagnet is a core.
Speed1	Speed refers to ratio of distance traveled to time interval.
Applied force1	We make a somewhat arbitrary distinction among the forces acting in a dynamical system. The categories are applied force, centering force and drag or friction force. The applied force is taken to be that force which is applied to the moving parts of a system by an outside agent as for example the force applied to a pendulum by someone pushing it or the force applied to a piece of metal by the magnetic field of an electromagnet. An applied force results in an energy transfer across the system boundary.
Force1	Force refers to agent that results in accelerating or deforming an object.
Displacement1	Displacement refers to change in position. A vector quantity.
Equilibrium1	Equilibrium refers to condition in which net force is equal to zero. Condition in which net torque on object is zero.
Position1	Separation between object and a reference point is a position.
Restoring force1	The force which tends to bring an oscillating body towards its mean position whenever it is displaced from the mean position is called the restoring force.
Natural frequency1	The frequency of vibration of an elastic object that depends on the size, composition, and shape of the object is referred to as natural frequency.
Decay1	Any process in which a particle disappears and in its place two or more different particles appear is decay.
Oscillation1	A to-and-fro or, side-to-side movement is an oscillation.

Chapter 6. TIME AND FREQUENCY CHARACTERIZATION OF SIGNALS AND SYSTEMS

Solid1	State of matter with fixed volume and shape is referred to as solid.
Unstable1	Matter that is capable of undergoing spontaneous change, as in a radioactive nuclide or an excited nuclear system. An unstable particle is any elementary particle that spontaneously decays into other particles.
Stable1	Does not decay is stable.
Converge1	If the dependent variable of a function under iteration gets closer and closer to a fixed value the function is said to converge under iteration.
Absorber1	Absorber refers to dark-coloured objects are good absorbers of infrared radiation. Light-coloured and silvered objects are poor absorbers.
Distance1	Distance refers to separation between two points. A scalar quantity.
Excitation1	The addition of energy to a system, transferring it from its ground state to an excited state. Excitation of a nucleus, an atom, or a molecule can result from absorption of photons or from inelastic collisions with other particles.
Pulse1	A wave of short duration confined to a small portion of the medium at any given time is called a pulse. A pulse is also called a wave pulse.
Period1	Period refers to time needed to repeat one complete cycle of motion.
Digital1	In electronics, meaning 'coded as numbers'. Digital means having discrete values as on a digital display.
Voltage1	Voltage refers to potential difference. It is a measure of the change in energy that one coulomb of electric charge undergoes when moved between 2 points.
Capacitor1	Electrical device used to store charge and energy in the electrical field is referred to as a capacitor.
Sine1	Sine refers to the ratio of the opposite side and the hypotenuse.
Electric motor1	Electric motor refers to a device consisting of a coil of wire and a magnet. The coil rotates when a current is in the coil.

Chapter 6. TIME AND FREQUENCY CHARACTERIZATION OF SIGNALS AND SYSTEMS

Angular velocity1	The rate of change of angular displacement is called angular velocity.
Velocity1	The ratio of change in position with respect to the time interval over which the change occurred is referred to as velocity.
State1	Dynamical systems evolve over the course of time. The state of the system at any instant may be identified by the values of certain variables at that instant. For example specifying the angle from the vertical and the velocity of a frictionless pendulum allows us to predict its position and velocity at any future time. Therefore the state of the pendulum at any instant is its position and velocity. In this example the position and velocity are known as state variables.
Modulation1	The process of superimposing information onto a carrier wave is referred to as modulation.
Noise1	Noise refers to sound consisting of an irregular mixture of frequencies.
Friction1	Force opposing relative motion of two objects are in contact is friction.
Element1	A pure substance that cannot be split up into anything simpler is called an element.

PRACTICE QUIZ
Chapter 6. TIME AND FREQUENCY CHARACTERIZATION OF SIGNALS AND SYSTEMS

1. In electronics, meaning 'coded as numbers'. _____ means having discrete values as on a _____ display.

    a.
    b.
    c. Digital
    d.

2. _____ refers to condition in which net force is equal to zero. Condition in which net torque on object is zero.

    a.
    b.
    c.
    d. Equilibrium

3. If the dependent variable of a function under iteration gets closer and closer to a fixed value the function is said to _____ under iteration.

    a.
    b.
    c. Converge
    d.

4. The _____ one uses should be of a size that makes sense for the particular subject at hand. It is easiest to define _____ in each area of science and then relate them to one another than to go around measuring particle masses in grams or cheese in proton mass units. In particle physics the standard unit is the unit of energy gev. One ev is the amount of energy that an electron gains when it moves through a potential difference of 1 Volt. G stands for Giga, or 10

    a. units
    b.
    c.
    d.

5. _____ refers to the size of a thing, without regard for its sign or direction. Similar to the absolute value of a number but applies to vectors as well.

----

a. Magnitude
b.
c.
d.

ANSWER KEY
Chapter 6. TIME AND FREQUENCY CHARACTERIZATION OF SIGNALS AND SYSTEMS

1. c
2. d
3. c
4. a

You can take the complete Chapter Practice Test

for Chapter 6. TIME AND FREQUENCY CHARACTERIZATION OF SIGNALS AND SYSTEMS
on all key terms, persons, places, and concepts.

Online 99 Cents

http://www.epub115.4.2097.6.cram101.com/

Use www.Cram101.com for all your study needs

including Cram101's online interactive problem solving labs in chemistry, statistics, mathematics, and more.

CHAPTER OUTLINE: KEY TERMS, PEOPLE, PLACES, CONCEPTS
Chapter 7
SAMPLING

- Property1
- Digital1
- System1
- Set1
- Frequency1
- Periodic1
- Impulse1
- Function1
- Period1
- Fundamental frequency1
- Domain1
- Magnitude1
- Phase1
- Solid1
- Equation1
- Distance1
- Normal1
- Slope1
- Second1

Chapter 7. SAMPLING

- Cycle1
- Amplitude1
- Speed1
- Rotation1
- Track1
- Position1
- Revolution1
- Harmonics1
- Conversion1
- Independent variable1
- Axis1
- Pulse1
- Properties1
- Range1
- Wave1
- Energy1
- Density1
- Light1
- Hertz1

Chapter 7. SAMPLING

Echo1

Receiver1

Generator1

Cosine1

Negative1

CHAPTER HIGHLIGHTS: KEY TERMS, PEOPLE, PLACES, CONCEPTS
Chapter 7. SAMPLING

Property1	A characteristic that is inherently associated with the object which is said to have that property. For example the mass of an object is one of its properties. So also might be color, density and many other characteristics. Properties are classified as extensive or intensive. Extensive properties increase in proportion to the size of the object, as mass does for example. Intensive properties are independent of the size of the object. The density for examples remains the same if I cut an object in half and throw half of it away. Things like an object's position or velocity are not considered to be properties of the object. They are not a characteristic of the object only but are also dependent on the reference frame in which the object is located.
Digital1	In electronics, meaning 'coded as numbers'. Digital means having discrete values as on a digital display.
System1	Defined collection of objects is called a system.
Set1	In mathematics a set is a collection of related objects. The mathematical usage is similar to the ordinary English meaning of the word. The objects that make up a set are called the elements of the set. If a set contains an unlimited number of elements it is an infinite set. Otherwise it is a finite set.
Frequency1	Frequency refers to number of occurrences per unit time.
Periodic1	A dynamical system or function which at some point returns to the same state or value. If a system or function ever revisits the identical state or value it will continue to come back to it again and again in equal intervals of time. That is why we call such a system periodic.
Impulse1	Product of force and time interval over which it acts is the impulse.
Function1	A mathematical function is a rule relating two sets of objects. Here we will restrict ourselves to objects that are numbers or vectors. One of the sets is called the domain of the function, the other is called the range of the function. Functions are frequently expressed as equations as for example Y=X+2. This function is interpreted as follows. For every X in the domain, add 2 to it to get the corresponding Y in the range. Because we are free to choose any X we want, X is called the independent variable. Because once X is chosen Y is fixed, we call Y the dependent variable.
Period1	Period refers to time needed to repeat one complete cycle of motion.
Fundamental frequency1	Fundamental frequency refers to the lowest frequency that can set up standing waves in an air column or on a string.

Chapter 7. SAMPLING

Domain1	The domain is the set of values that the independent variable of a function may take on. A domain may be finite as in the set of numbers {1, 2, 3.n} or infinite as in all the numbers between 0 and 1.
Magnitude1	Magnitude refers to the size of a thing, without regard for its sign or direction. Similar to the absolute value of a number but applies to vectors as well.
Phase1	The particles in a wave, which are in the same state of vibration, i.e., the same position and the same direction of motion are said to be in the same phase.
Solid1	State of matter with fixed volume and shape is referred to as solid.
Equation1	An equation is a mathematical expression with an equal sign in it. It signifies that the numerical or vector value on one side of the = is the same as the numerical or vector value on the other side. An equation may include variables and parameters. If any of the variables are rates of change, the equation is called a differential equation.
Distance1	Distance refers to separation between two points. A scalar quantity.
Normal1	Perpendicular to plane of interest is called normal.
Slope1	Ratio of the vertical separation, or rise to the horizontal separation, or run are called the slope.
Second1	Second is a SI unit of time.
Cycle1	In wave motion, one cycle is a trough and a crest for a transverse wave, or a compression and a rarefaction for a longitudinal wave.
Amplitude1	In any periodic motion, the maximum displacement from equilibrium is called amplitude.
Speed1	Speed refers to ratio of distance traveled to time interval.
Rotation1	Rotation refers to the spin of an object around its central axis. Earth rotates about its axis every 24 hours. A spinning top rotates about its center shaft.
Track1	The record of the path of a particle traversing a detector is a track.
Position1	Separation between object and a reference point is a position.

Chapter 7. SAMPLING

Revolution1	Revolution refers to the orbital motion of one object around another. The Earth revolves around the Sun in one year. The moon revolves around the Earth in approximately 28 days.
Harmonics1	Frequencies produced by musical instrument that are multiples of fundamental tone are referred to as harmonics.
Conversion1	Chemical process turning U308 into UF6 preparatory to enrichment is conversion.
Independent variable1	Variable that is manipulated or changed in an experiment is referred to as the independent variable.
Axis1	Axis refers to the imaginary line about which a planet or other object rotates .
Pulse1	A wave of short duration confined to a small portion of the medium at any given time is called a pulse. A pulse is also called a wave pulse.
Properties1	Properties refers to qualities or attributes that, taken together, are usually unique to an object; for example, color, texture, and size.
Range1	The range is the set of values that the dependent variable of a function may take on. A range may be finite as in the set of numbers {1, 2, 3.n} or infinite as in all the mumbers between 0 and 1.
Wave1	A set of oscillations or vibrations that transfer energy without any transfer of mass is the wave.
Energy1	Energy refers to non-material property capable of causing changes in matter.
Density1	The mass of a given volume of substance. It has units of kg/m3 or g/cm3. When the density is high the particles are closely packed.
Light1	Electromagnetic radiation with wavelengths between 400 and 700 nm that is visible is called light.
Hertz1	Unit of frequency equal to one event or cycle per second is called hertz.
Echo1	A reflection of a sound or ultrasound is called echo.
Receiver1	Device that detects electromagnetic waves is called receiver.

Chapter 7. SAMPLING

Generator1 | Generator produces electricity when a magnet spins inside a coil of wire or a coil of wire spins inside a magnetic field.

Cosine1 | The ratio of the adjacent side to the hypotenuse is the cosine.

Negative1 | The sign of the electric charge on the electron is negative.

PRACTICE QUIZ
Chapter 7. SAMPLING

1. State of matter with fixed volume and shape is referred to as _____.

    a.
    b.
    c.
    d. Solid

2. The particles in a wave, which are in the same state of vibration, i.e., the same position and the same direction of motion are said to be in the same _____.

    a.
    b.
    c. Phase
    d.

3. Frequencies produced by musical instrument that are multiples of fundamental tone are referred to as _____.

    a.
    b.
    c.
    d. Harmonics

4. A dynamical system or function which at some point returns to the same state or value. If a system or function ever revisits the identical state or value it will continue to come back to it again and again in equal intervals of time. That is why we call such a system _____.

    a.
    b.
    c.
    d. Periodic

5. In electronics, meaning 'coded as numbers'. _____ means having discrete values as on a _____ display.

    a.
    b.
    c. Digital
    d.

ANSWER KEY
Chapter 7. SAMPLING

1. d
2. c
3. d
4. d
5. c

You can take the complete Chapter Practice Test

for Chapter 7. SAMPLING
on all key terms, persons, places, and concepts.

Online 99 Cents

http://www.epub115.4.2097.7.cram101.com/

Use www.Cram101.com for all your study needs

including Cram101's online interactive problem solving labs in chemistry, statistics, mathematics, and more.

CHAPTER OUTLINE: KEY TERMS, PEOPLE, PLACES, CONCEPTS
Chapter 8
COMMUNICATION SYSTEMS

- Receiver1
- Frequency1
- Range1
- Position1
- Second1
- Modulation1
- Amplitude1
- Microwave1
- Satellite1
- Property1
- Cow1
- Domain1
- Axis1
- Phase1
- System1
- Magnitude1
- Detector1
- Power1
- Current1

Chapter 8. COMMUNICATION SYSTEMS

- Negative1
- Attenuation1
- Wave1
- Stable1
- Pulse1
- Reflection1
- Radar1
- Ionosphere1
- Noise1
- Vapor1
- Absorption1
- Infrared1
- Light1
- Ultraviolet1
- Energy1
- Conversion1
- Efficiency1
- Set1
- Periodic1

Chapter 8. COMMUNICATION SYSTEMS

- Period 1
- Impulse 1
- Interference 1
- Proportionality constant 1
- Dispersion 1
- Accuracy 1
- Symmetry 1
- Digital 1
- Quantized 1
- Properties 1
- Function 1
- Equation 1
- Fundamental frequency 1
- Excursion 1
- Decay 1
- Track 1
- Cosine 1
- Normal 1
- Absolute value 1

Chapter 8. COMMUNICATION SYSTEMS

Random1

Element1

Diode1

Voltage1

Time interval1

State1

CHAPTER HIGHLIGHTS: KEY TERMS, PEOPLE, PLACES, CONCEPTS
Chapter 8. COMMUNICATION SYSTEMS

Receiver1	Device that detects electromagnetic waves is called receiver.
Frequency1	Frequency refers to number of occurrences per unit time.
Range1	The range is the set of values that the dependent variable of a function may take on. A range may be finite as in the set of numbers {1, 2, 3.n} or infinite as in all the mumbers between 0 and 1.
Position1	Separation between object and a reference point is a position.
Second1	Second is a SI unit of time.
Modulation1	The process of superimposing information onto a carrier wave is referred to as modulation.
Amplitude1	In any periodic motion, the maximum displacement from equilibrium is called amplitude.
Microwave1	Short wavelength electromagnetic waves used for communications and cooking are called microwave.
Satellite1	An object that orbits the Earth or other astronomical body is the satellite.
Property1	A characteristic that is inherently associated with the object which is said to have that property. For example the mass of an object is one of its properties. So also might be color, density and many other characteristics. Properties are classified as extensive or intensive. Extensive properties increase in proportion to the size of the object, as mass does for example. Intensive properties are independent of the size of the object. The density for examples remains the same if I cut an object in half and throw half of it away. Things like an object's position or velocity are not considered to be properties of the object. They are not a characteristic of the object only but are also dependent on the reference frame in which the object is located.
Cow1	A radioisotope generator system is referred to as cow.
Domain1	The domain is the set of values that the independent variable of a function may take on. A domain may be finite as in the set of numbers {1, 2, 3.n} or infinite as in all the numbers between 0 and 1.
Axis1	Axis refers to the imaginary line about which a planet or other object rotates .

Chapter 8. COMMUNICATION SYSTEMS

Phase1	The particles in a wave, which are in the same state of vibration, i.e., the same position and the same direction of motion are said to be in the same phase.
System1	Defined collection of objects is called a system.
Magnitude1	Magnitude refers to the size of a thing, without regard for its sign or direction. Similar to the absolute value of a number but applies to vectors as well.
Detector1	Detector refers to any device used to sense the passage of a particle; also a collection of such devices designed so that each serves a particular purpose in allowing physicists to reconstruct particle events.
Power1	Power refers to rate of doing work; rate of energy conversion.
Current1	Current refers to a flow of charge. Measured in amps.
Negative1	The sign of the electric charge on the electron is negative.
Attenuation1	The process by which a compound is reduced in concentration over time, through adsorption, degradation, dilution, and/or transformation. Radiologically, it is the reduction of the intensity of radiation upon passage through a medium. The attenuation is caused by absorption and scattering.
Wave1	A set of oscillations or vibrations that transfer energy without any transfer of mass is the wave.
Stable1	Does not decay is stable.
Pulse1	A wave of short duration confined to a small portion of the medium at any given time is called a pulse. A pulse is also called a wave pulse.
Reflection1	Reflection refers to the change when light, sound, or other waves bounce backwards off a boundary .
Radar1	A method of detecting, locating, or tracking an object by using beamed, reflected, and timed radio waves. RADAR also refers to the electronic equipment that uses radio waves to detect, locate, and track objects.

Chapter 8. COMMUNICATION SYSTEMS

Ionosphere1	An ionosphere is a region covering the highest layers in the Earth's atmosphere, containing an appreciable population of ions and free electrons. The ions are created by sunlight ranging from the ultra-violet to x-rays. In the lowest and least rarefied layer of the ionosphere, the D-layer, as soon as the Sun sets the ions and electrons recombine, but in the higher layers, collisions are so few that its ion layers last throughout the night.
Noise1	Noise refers to sound consisting of an irregular mixture of frequencies.
Vapor1	The gaseous state of a substance that is normally in the liquid state is vapor.
Absorption1	The transfer of energy to a medium, such as body tissues, as a radiation beam passes through the medium is absorption.
Infrared1	A type of electromagnetic radiation with a wavelength longer than that of light is called infrared.
Light1	Electromagnetic radiation with wavelengths between 400 and 700 nm that is visible is called light.
Ultraviolet1	Ultraviolet refers to electromagnetic waves with a wavelength shorter than that of visible light.
Energy1	Energy refers to non-material property capable of causing changes in matter.
Conversion1	Chemical process turning U308 into UF6 preparatory to enrichment is conversion.
Efficiency1	Ratio of output work to input work is called efficiency.
Set1	In mathematics a set is a collection of related objects. The mathematical usage is similar to the ordinary English meaning of the word. The objects that make up a set are called the elements of the set. If a set contains an unlimited number of elements it is an infinite set. Otherwise it is a finite set.
Periodic1	A dynamical system or function which at some point returns to the same state or value. If a system or function ever revisits the identical state or value it will continue to come back to it again and again in equal intervals of time. That is why we call such a system periodic.
Period1	Period refers to time needed to repeat one complete cycle of motion.
Impulse1	Product of force and time interval over which it acts is the impulse.

Chapter 8. COMMUNICATION SYSTEMS

Interference1	Phenomenon of light where the relative phase difference between two light waves produces light or dark spots, a result of light's wavelike nature is called the interference.
Proportionality constant1	A constant applied to a proportionality statement that transforms the statement into an equation is a proportionality constant.
Dispersion1	The splitting of light into its constituent colours is called dispersion.
Accuracy1	Accuracy refers to closeness of a measurement to the standard value of that quantity.
Symmetry1	Symmetry refers to property that is now charged when operation or reference frame is charged.
Digital1	In electronics, meaning 'coded as numbers'. Digital means having discrete values as on a digital display.
Quantized1	Quantized refers to a quantity that cannot be divided into smaller increments forever, for which there exists a minimum, quantum increment.
Properties1	Properties refers to qualities or attributes that, taken together, are usually unique to an object; for example, color, texture, and size.
Function1	A mathematical function is a rule relating two sets of objects. Here we will restrict ourselves to objects that are numbers or vectors. One of the sets is called the domain of the function, the other is called the range of the function. Functions are frequently expressed as equations as for example Y=X+2. This function is interpreted as follows. For every X in the domain, add 2 to it to get the corresponding Y in the range. Because we are free to choose any X we want, X is called the independent variable. Because once X is chosen Y is fixed, we call Y the dependent variable.
Equation1	An equation is a mathematical expression with an equal sign in it. It signifies that the numerical or vector value on one side of the = is the same as the numerical or vector value on the other side. An equation may include variables and parameters. If any of the variables are rates of change, the equation is called a differential equation.
Fundamental frequency1	Fundamental frequency refers to the lowest frequency that can set up standing waves in an air column or on a string .

Chapter 8. COMMUNICATION SYSTEMS

Excursion1	Excursion refers to where the leaching solutions used in the uranium In Situ Leaching mining technique escape outside the mining zone.
Decay1	Any process in which a particle disappears and in its place two or more different particles appear is decay.
Track1	The record of the path of a particle traversing a detector is a track.
Cosine1	The ratio of the adjacent side to the hypotenuse is the cosine.
Normal1	Perpendicular to plane of interest is called normal.
Absolute value1	The absolute value of a number is -1 times the number if it is negative or +1 times the number if it is positive. In modern statistics this is accomplished by taking the square root of the number squared.
Random1	With no set order or pattern, we have random.
Element1	A pure substance that cannot be split up into anything simpler is called an element.
Diode1	Electrical device permitting only one way current flow is a diode.
Voltage1	Voltage refers to potential difference. It is a measure of the change in energy that one coulomb of electric charge undergoes when moved between 2 points.
Time interval1	Difference in time between two clock readings is referred to as a time interval.
State1	Dynamical systems evolve over the course of time. The state of the system at any instant may be identified by the values of certain variables at that instant. For example specifying the angle from the vertical and the velocity of a frictionless pendulum allows us to predict its position and velocity at any future time. Therefore the state of the pendulum at any instant is its position and velocity. In this example the position and velocity are known as state variables.

PRACTICE QUIZ
Chapter 8. COMMUNICATION SYSTEMS

1. With no set order or pattern, we have _____.

    a.
    b.
    c.
    d. Random

2. _____ refers to the lowest frequency that can set up standing waves in an air column or on a string.

    a.
    b.
    c.
    d. Fundamental frequency

3. A wave of short duration confined to a small portion of the medium at any given time is called a pulse. A _____ is also called a wave pulse.

    a.
    b.
    c.
    d. Pulse

4. _____ is a SI unit of time.

    a.
    b.
    c.
    d. Second

5. The ratio of the adjacent side to the hypotenuse is the _____.

    a.
    b.
    c. Cosine
    d.

ANSWER KEY
Chapter 8. COMMUNICATION SYSTEMS

1. d
2. d
3. d
4. d
5. c

You can take the complete Chapter Practice Test

for Chapter 8. COMMUNICATION SYSTEMS
on all key terms, persons, places, and concepts.

Online 99 Cents

http://www.epub115.4.2097.8.cram101.com/

Use www.Cram101.com for all your study needs

including Cram101's online interactive problem solving labs in chemistry, statistics, mathematics, and more.

CHAPTER OUTLINE: KEY TERMS, PEOPLE, PLACES, CONCEPTS
Chapter 9
THE LAPLACE TRANSFORM

- Periodic1
- Property1
- Properties1
- Unstable1
- Set1
- System1
- Impulse1
- Function1
- Independent variable1
- Negative1
- Converge1
- Second1
- Range1
- Complex number1
- Complex plane1
- Axis1
- Poles1
- Domain1
- Equation1

Chapter 9. THE LAPLACE TRANSFORM

- Vector1
- Magnitude1
- Frequency1
- Phase1
- Slope1
- Speed1
- Distance1
- Origin1
- Position1
- Movement1
- Decay1
- Natural frequency1
- Attenuation1
- Current1
- Compression1
- Stable1
- Capacitor1
- Voltage1
- Resistor1

Chapter 9. THE LAPLACE TRANSFORM

- Power1
- Diverge1
- Symmetry1
- Singularity1
- State1
- Pressure1
- Echo1
- Receiver1
- Amplitude1
- Volt1
- Switch1

CHAPTER HIGHLIGHTS: KEY TERMS, PEOPLE, PLACES, CONCEPTS
Chapter 9. THE LAPLACE TRANSFORM

Periodic1	A dynamical system or function which at some point returns to the same state or value. If a system or function ever revisits the identical state or value it will continue to come back to it again and again in equal intervals of time. That is why we call such a system periodic.
Property1	A characteristic that is inherently associated with the object which is said to have that property. For example the mass of an object is one of its properties. So also might be color, density and many other characteristics. Properties are classified as extensive or intensive. Extensive properties increase in proportion to the size of the object, as mass does for example. Intensive properties are independent of the size of the object. The density for examples remains the same if I cut an object in half and throw half of it away. Things like an object's position or velocity are not considered to be properties of the object. They are not a characteristic of the object only but are also dependent on the reference frame in which the object is located.
Properties1	Properties refers to qualities or attributes that, taken together, are usually unique to an object; for example, color, texture, and size.
Unstable1	Matter that is capable of undergoing spontaneous change, as in a radioactive nuclide or an excited nuclear system. An unstable particle is any elementary particle that spontaneously decays into other particles.
Set1	In mathematics a set is a collection of related objects. The mathematical usage is similar to the ordinary English meaning of the word. The objects that make up a set are called the elements of the set. If a set contains an unlimited number of elements it is an infinite set. Otherwise it is a finite set.
System1	Defined collection of objects is called a system.
Impulse1	Product of force and time interval over which it acts is the impulse.
Function1	A mathematical function is a rule relating two sets of objects. Here we will restrict ourselves to objects that are numbers or vectors. One of the sets is called the domain of the function, the other is called the range of the function. Functions are frequently expressed as equations as for example Y=X+2. This function is interpreted as follows. For every X in the domain, add 2 to it to get the corresponding Y in the range. Because we are free to choose any X we want, X is called the independent variable. Because once X is chosen Y is fixed, we call Y the dependent variable.
Independent variable1	Variable that is manipulated or changed in an experiment is referred to as the independent variable.

Chapter 9. THE LAPLACE TRANSFORM

Negative1	The sign of the electric charge on the electron is negative.
Converge1	If the dependent variable of a function under iteration gets closer and closer to a fixed value the function is said to converge under iteration.
Second1	Second is a SI unit of time.
Range1	The range is the set of values that the dependent variable of a function may take on. A range may be finite as in the set of numbers {1, 2, 3.n} or infinite as in all the mumbers between 0 and 1.
Complex number1	Ordinary numbers that we use for accounting and simple calculations are called real numbers. There is no real number such that the square of that number is -1 since the product of any real number with itself is positive. To remedy this situation, the square root of -1 was defined and given the symbol i. Then a new class of numbers was invented, called imaginary numbers, made up of i times the set of real numbers. A complex number is a number comprised of a real part added to an imaginary part like a+b*i.
Complex plane1	The complex plane contains the set of complex numbers. It is a plane spanned by the set of real numbers, normally along the horizontal axis, and the set of imaginary numbers, normally along the vertical axis.
Axis1	Axis refers to the imaginary line about which a planet or other object rotates .
Poles1	The parts of a magnet where the magnetic field is strongest are poles.
Domain1	The domain is the set of values that the independent variable of a function may take on. A domain may be finite as in the set of numbers {1, 2, 3.n} or infinite as in all the numbers between 0 and 1.
Equation1	An equation is a mathematical expression with an equal sign in it. It signifies that the numerical or vector value on one side of the = is the same as the numerical or vector value on the other side. An equation may include variables and parameters. If any of the variables are rates of change, the equation is called a differential equation.
Vector1	Any quantity that has both magnitude and direction. Velocity is a vector.
Magnitude1	Magnitude refers to the size of a thing, without regard for its sign or direction. Similar to the absolute value of a number but applies to vectors as well.

Chapter 9. THE LAPLACE TRANSFORM

Frequency1	Frequency refers to number of occurrences per unit time.
Phase1	The particles in a wave, which are in the same state of vibration, i.e., the same position and the same direction of motion are said to be in the same phase.
Slope1	Ratio of the vertical separation, or rise to the horizontal separation, or run are called the slope.
Speed1	Speed refers to ratio of distance traveled to time interval.
Distance1	Distance refers to separation between two points. A scalar quantity.
Origin1	Origin refers to the point in a reference frame from which measurements are made. It is the location of the zero value for each axis in the frame.
Position1	Separation between object and a reference point is a position.
Movement1	Change of position is called movement.
Decay1	Any process in which a particle disappears and in its place two or more different particles appear is decay.
Natural frequency1	The frequency of vibration of an elastic object that depends on the size, composition, and shape of the object is referred to as natural frequency.
Attenuation1	The process by which a compound is reduced in concentration over time, through adsorption, degradation, dilution, and/or transformation. Radiologically, it is the reduction of the intensity of radiation upon passage through a medium. The attenuation is caused by absorption and scattering.
Current1	Current refers to a flow of charge. Measured in amps.
Compression1	In a sound wave, a compression is a region in the material that transmits the wave where the particles are closer together than normal.
Stable1	Does not decay is stable.
Capacitor1	Electrical device used to store charge and energy in the electrical field is referred to as a capacitor.

Chapter 9. THE LAPLACE TRANSFORM

Voltage1	Voltage refers to potential difference. It is a measure of the change in energy that one coulomb of electric charge undergoes when moved between 2 points.
Resistor1	Device designed to have a specific resistance is a resistor.
Power1	Power refers to rate of doing work; rate of energy conversion.
Diverge1	If the dependent variable of a function under iteration increases without limit the function is said to diverge under iteration.
Symmetry1	Symmetry refers to property that is now charged when operation or reference frame is charged.
Singularity1	The center of a black hole, where the curvature of spacetime is maximal is its singularity. At the singularity, the gravitational tides diverge; no solid object can even theoretically survive hitting the singularity. Although singularities generally predict inconsistencies in theory, singularities within black holes do not necessarily imply that general relativity is incomplete so long as singularities are always surrounded by event horizons. A proper formulation of quantum gravity may well avoid the classical singularity at the centers of black holes.
State1	Dynamical systems evolve over the course of time. The state of the system at any instant may be identified by the values of certain variables at that instant. For example specifying the angle from the vertical and the velocity of a frictionless pendulum allows us to predict its position and velocity at any future time. Therefore the state of the pendulum at any instant is its position and velocity. In this example the position and velocity are known as state variables.
Pressure1	Force per unit area is referred to as the pressure.
Echo1	A reflection of a sound or ultrasound is called echo.
Receiver1	Device that detects electromagnetic waves is called receiver.
Amplitude1	In any periodic motion, the maximum displacement from equilibrium is called amplitude.
Volt1	The unit of voltage or potential difference is a volt.
Switch1	Switch refers to a device in an electrical circuit that breaks or makes a complete path for the current.

PRACTICE QUIZ
Chapter 9. THE LAPLACE TRANSFORM

1. Defined collection of objects is called a _____.

    a.
    b. System
    c.
    d.

2. The _____ is the set of values that the independent variable of a function may take on. A _____ may be finite as in the set of numbers {1, 2, 3.n} or infinite as in all the numbers between 0 and 1.

    a.
    b.
    c.
    d. Domain

3. _____ refers to number of occurrences per unit time.

    a. Frequency
    b.
    c.
    d.

4. A mathematical _____ is a rule relating two sets of objects. Here we will restrict ourselves to objects that are numbers or vectors. One of the sets is called the domain of the function, the other is called the range of the function. Functions are frequently expressed as equations as for example Y=X+2. This _____ is interpreted as follows. For every X in the domain, add 2 to it to get the corresponding Y in the range. Because we are free to choose any X we want, X is called the independent variable. Because once X is chosen Y is fixed, we call Y the dependent variable.

    a.
    b.
    c.
    d. Function

5. _____ refers to the point in a reference frame from which measurements are made. It is the location of the zero value for each axis in the frame.

----

a.
b. Origin
c.
d.

ANSWER KEY
Chapter 9. THE LAPLACE TRANSFORM

1. b
2. d
3. a
4. d
5. b

You can take the complete Chapter Practice Test

for Chapter 9. THE LAPLACE TRANSFORM
on all key terms, persons, places, and concepts.

Online 99 Cents

http://www.epub115.4.2097.9.cram101.com/

Use www.Cram101.com for all your study needs

including Cram101's online interactive problem solving labs in chemistry, statistics, mathematics, and more.

CHAPTER OUTLINE: KEY TERMS, PEOPLE, PLACES, CONCEPTS
Chapter 10
THE Z - TRANSFORM

- Converge1
- Unstable1
- System1
- Properties1
- Impulse1
- Magnitude1
- Axis1
- Range1
- Function1
- Poles1
- Property1
- Origin1
- Equation1
- Complex plane1
- Negative1
- Boundary1
- Decay1
- Proof1
- Power1

Chapter 10. THE Z - TRANSFORM

- Frequency1
- Vector1
- Phase1
- Speed1
- Distance1
- Rotation1
- Set1
- Domain1
- Amplitude1
- Stable1
- Second1
- Element1
- Accuracy1
- Digital1
- State1
- Complex number1

CHAPTER HIGHLIGHTS: KEY TERMS, PEOPLE, PLACES, CONCEPTS
Chapter 10. THE Z - TRANSFORM

Converge1	If the dependent variable of a function under iteration gets closer and closer to a fixed value the function is said to converge under iteration.
Unstable1	Matter that is capable of undergoing spontaneous change, as in a radioactive nuclide or an excited nuclear system. An unstable particle is any elementary particle that spontaneously decays into other particles.
System1	Defined collection of objects is called a system.
Properties1	Properties refers to qualities or attributes that, taken together, are usually unique to an object; for example, color, texture, and size.
Impulse1	Product of force and time interval over which it acts is the impulse.
Magnitude1	Magnitude refers to the size of a thing, without regard for its sign or direction. Similar to the absolute value of a number but applies to vectors as well.
Axis1	Axis refers to the imaginary line about which a planet or other object rotates .
Range1	The range is the set of values that the dependent variable of a function may take on. A range may be finite as in the set of numbers {1, 2, 3.n} or infinite as in all the mumbers between 0 and 1.
Function1	A mathematical function is a rule relating two sets of objects. Here we will restrict ourselves to objects that are numbers or vectors. One of the sets is called the domain of the function, the other is called the range of the function. Functions are frequently expressed as equations as for example Y=X+2. This function is interpreted as follows. For every X in the domain, add 2 to it to get the corresponding Y in the range. Because we are free to choose any X we want, X is called the independent variable. Because once X is chosen Y is fixed, we call Y the dependent variable.
Poles1	The parts of a magnet where the magnetic field is strongest are poles.

Chapter 10. THE Z - TRANSFORM

Property1	A characteristic that is inherently associated with the object which is said to have that property. For example the mass of an object is one of its properties. So also might be color, density and many other characteristics. Properties are classified as extensive or intensive. Extensive properties increase in proportion to the size of the object, as mass does for example. Intensive properties are independent of the size of the object. The density for examples remains the same if I cut an object in half and throw half of it away. Things like an object's position or velocity are not considered to be properties of the object. They are not a characteristic of the object only but are also dependent on the reference frame in which the object is located.
Origin1	Origin refers to the point in a reference frame from which measurements are made. It is the location of the zero value for each axis in the frame.
Equation1	An equation is a mathematical expression with an equal sign in it. It signifies that the numerical or vector value on one side of the = is the same as the numerical or vector value on the other side. An equation may include variables and parameters. If any of the variables are rates of change, the equation is called a differential equation.
Complex plane1	The complex plane contains the set of complex numbers. It is a plane spanned by the set of real numbers, normally along the horizontal axis, and the set of imaginary numbers, normally along the vertical axis.
Negative1	The sign of the electric charge on the electron is negative.
Boundary1	Boundary refers to the division between two regions of differing physical properties .
Decay1	Any process in which a particle disappears and in its place two or more different particles appear is decay.
Proof1	A measure of ethanol concentration of an alcoholic beverage; proof is double the concentration by volume; for example, 50 percent by volume is 100 proof.
Power1	Power refers to rate of doing work; rate of energy conversion.
Frequency1	Frequency refers to number of occurrences per unit time.
Vector1	Any quantity that has both magnitude and direction. Velocity is a vector.
Phase1	The particles in a wave, which are in the same state of vibration, i.e., the same position and the same direction of motion are said to be in the same phase.

Chapter 10. THE Z - TRANSFORM

Speed1	Speed refers to ratio of distance traveled to time interval.
Distance1	Distance refers to separation between two points. A scalar quantity.
Rotation1	Rotation refers to the spin of an object around its central axis. Earth rotates about its axis every 24 hours. A spinning top rotates about its center shaft.
Set1	In mathematics a set is a collection of related objects. The mathematical usage is similar to the ordinary English meaning of the word. The objects that make up a set are called the elements of the set. If a set contains an unlimited number of elements it is an infinite set. Otherwise it is a finite set.
Domain1	The domain is the set of values that the independent variable of a function may take on. A domain may be finite as in the set of numbers {1, 2, 3.n} or infinite as in all the numbers between 0 and 1.
Amplitude1	In any periodic motion, the maximum displacement from equilibrium is called amplitude.
Stable1	Does not decay is stable.
Second1	Second is a SI unit of time.
Element1	A pure substance that cannot be split up into anything simpler is called an element.
Accuracy1	Accuracy refers to closeness of a measurement to the standard value of that quantity.
Digital1	In electronics, meaning 'coded as numbers'. Digital means having discrete values as on a digital display.
State1	Dynamical systems evolve over the course of time. The state of the system at any instant may be identified by the values of certain variables at that instant. For example specifying the angle from the vertical and the velocity of a frictionless pendulum allows us to predict its position and velocity at any future time. Therefore the state of the pendulum at any instant is its position and velocity. In this example the position and velocity are known as state variables.

Chapter 10. THE Z - TRANSFORM

Complex number1	Ordinary numbers that we use for accounting and simple calculations are called real numbers. There is no real number such that the square of that number is -1 since the product of any real number with itself is positive. To remedy this situation, the square root of -1 was defined and given the symbol i. Then a new class of numbers was invented, called imaginary numbers, made up of i times the set of real numbers. A complex number is a number comprised of a real part added to an imaginary part like a+b*i.

PRACTICE QUIZ
Chapter 10. THE Z - TRANSFORM

1. _____ refers to separation between two points. A scalar quantity.

   a.
   b.
   c.
   d. Distance

2. Does not decay is _____.

   a.
   b.
   c. Stable
   d.

3. Matter that is capable of undergoing spontaneous change, as in a radioactive nuclide or an excited nuclear system. An _____ particle is any elementary particle that spontaneously decays into other particles.

   a.
   b.
   c. Unstable
   d.

4. _____ refers to the division between two regions of differing physical properties.

   a.
   b.
   c.
   d. Boundary

5. Product of force and time interval over which it acts is the _____.

   a. Impulse
   b.
   c.
   d.

ANSWER KEY
Chapter 10. THE Z - TRANSFORM

1. d
2. c
3. c
4. d
5. a

You can take the complete Chapter Practice Test

for Chapter 10. THE Z - TRANSFORM
on all key terms, persons, places, and concepts.

Online 99 Cents

http://www.epub115.4.2097.10.cram101.com/

Use www.Cram101.com for all your study needs

including Cram101's online interactive problem solving labs in chemistry, statistics, mathematics, and more.

CHAPTER OUTLINE: KEY TERMS, PEOPLE, PLACES, CONCEPTS
Chapter 11
LINEAR FEEDBACK SYSTEMS

- System1
- Position1
- Voltage1
- Potentiometer1
- Second1
- Electric motor1
- Object1
- Distance1
- Velocity1
- Unstable1
- Weight1
- Track1
- Trajectory1
- Movement1
- Function1
- Properties1
- Range1
- Equation1
- Capacitor1

Chapter 11. LINEAR FEEDBACK SYSTEMS

- Property1
- Current1
- Diode1
- Frequency1
- Vacuum1
- Magnitude1
- Temperature1
- Phase1
- Stable1
- Negative1
- Power1
- Poles1
- Axis1
- Impulse1
- Dynamics1
- Restoring force1
- Force1
- Proportion1
- Digital1

Chapter 11. LINEAR FEEDBACK SYSTEMS

- Conversion1
- Element1
- Noise1
- Energy1
- Sound1
- Speed1
- Attenuation1
- Complex plane1
- Set1
- Quantity1
- Symmetry1
- Accuracy1
- Vector1
- Origin1
- State1
- Reflection1
- Revolution1
- Oscillation1
- Nominal1

Chapter 11. LINEAR FEEDBACK SYSTEMS

- Precision 1
- Natural frequency 1
- Units 1
- Rad 1
- Gas 1
- Converge 1
- Singularity 1
- Gravity 1
- Charge 1
- Electron 1
- Volt 1
- Resistor 1
- Resistance 1
- Capacitance 1
- Acceleration 1
- Reference point 1
- Angular acceleration 1
- Equilibrium 1

CHAPTER HIGHLIGHTS: KEY TERMS, PEOPLE, PLACES, CONCEPTS
Chapter 11. LINEAR FEEDBACK SYSTEMS

System1	Defined collection of objects is called a system.
Position1	Separation between object and a reference point is a position.
Voltage1	Voltage refers to potential difference. It is a measure of the change in energy that one coulomb of electric charge undergoes when moved between 2 points.
Potentiometer1	Potentiometer refers to electrical device with variable resistance; rheostat.
Second1	Second is a SI unit of time.
Electric motor1	Electric motor refers to a device consisting of a coil of wire and a magnet. The coil rotates when a current is in the coil.
Object1	Object refers to source of diverging light rays; either luminous or illuminated.
Distance1	Distance refers to separation between two points. A scalar quantity.
Velocity1	The ratio of change in position with respect to the time interval over which the change occurred is referred to as velocity.
Unstable1	Matter that is capable of undergoing spontaneous change, as in a radioactive nuclide or an excited nuclear system. An unstable particle is any elementary particle that spontaneously decays into other particles.
Weight1	Force of gravity of an object is called weight.
Track1	The record of the path of a particle traversing a detector is a track.
Trajectory1	The path followed by a projectile is referred to as trajectory.
Movement1	Change of position is called movement.

Chapter 11. LINEAR FEEDBACK SYSTEMS

Function1	A mathematical function is a rule relating two sets of objects. Here we will restrict ourselves to objects that are numbers or vectors. One of the sets is called the domain of the function, the other is called the range of the function. Functions are frequently expressed as equations as for example Y=X+2. This function is interpreted as follows. For every X in the domain, add 2 to it to get the corresponding Y in the range. Because we are free to choose any X we want, X is called the independent variable. Because once X is chosen Y is fixed, we call Y the dependent variable.
Properties1	Properties refers to qualities or attributes that, taken together, are usually unique to an object; for example, color, texture, and size.
Range1	The range is the set of values that the dependent variable of a function may take on. A range may be finite as in the set of numbers {1, 2, 3.n} or infinite as in all the mumbers between 0 and 1.
Equation1	An equation is a mathematical expression with an equal sign in it. It signifies that the numerical or vector value on one side of the = is the same as the numerical or vector value on the other side. An equation may include variables and parameters. If any of the variables are rates of change, the equation is called a differential equation.
Capacitor1	Electrical device used to store charge and energy in the electrical field is referred to as a capacitor.
Property1	A characteristic that is inherently associated with the object which is said to have that property. For example the mass of an object is one of its properties. So also might be color, density and many other characteristics. Properties are classified as extensive or intensive. Extensive properties increase in proportion to the size of the object, as mass does for example. Intensive properties are independent of the size of the object. The density for examples remains the same if I cut an object in half and throw half of it away. Things like an object's position or velocity are not considered to be properties of the object. They are not a characteristic of the object only but are also dependent on the reference frame in which the object is located.
Current1	Current refers to a flow of charge. Measured in amps.
Diode1	Electrical device permitting only one way current flow is a diode.
Frequency1	Frequency refers to number of occurrences per unit time.
Vacuum1	Vacuum refers to a region of space containing no matter. In practice, a region of gas at very low pressure.

Chapter 11. LINEAR FEEDBACK SYSTEMS

Magnitude1	Magnitude refers to the size of a thing, without regard for its sign or direction. Similar to the absolute value of a number but applies to vectors as well.
Temperature1	Temperature refers to measure of hotness of object on a quantitative scale. In gases, proportional to average kinetic energy of molecules.
Phase1	The particles in a wave, which are in the same state of vibration, i.e., the same position and the same direction of motion are said to be in the same phase.
Stable1	Does not decay is stable.
Negative1	The sign of the electric charge on the electron is negative.
Power1	Power refers to rate of doing work; rate of energy conversion.
Poles1	The parts of a magnet where the magnetic field is strongest are poles.
Axis1	Axis refers to the imaginary line about which a planet or other object rotates.
Impulse1	Product of force and time interval over which it acts is the impulse.
Dynamics1	Study of motion of particles acted on by forces is called dynamics.
Restoring force1	The force which tends to bring an oscillating body towards its mean position whenever it is displaced from the mean position is called the restoring force.
Force1	Force refers to agent that results in accelerating or deforming an object.
Proportion1	Proportion refers to two quantities are directly proportional if doubling one of them has the effect of doubling the other. On a graph we get a straight line through the origin.
Digital1	In electronics, meaning 'coded as numbers'. Digital means having discrete values as on a digital display.
Conversion1	Chemical process turning U308 into UF6 preparatory to enrichment is conversion.
Element1	A pure substance that cannot be split up into anything simpler is called an element.

Chapter 11. LINEAR FEEDBACK SYSTEMS

Noise1	Noise refers to sound consisting of an irregular mixture of frequencies.
Energy1	Energy refers to non-material property capable of causing changes in matter.
Sound1	A longitudinal wave, usually through the air but can travel through liquids and solids is called sound.
Speed1	Speed refers to ratio of distance traveled to time interval.
Attenuation1	The process by which a compound is reduced in concentration over time, through adsorption, degradation, dilution, and/or transformation. Radiologically, it is the reduction of the intensity of radiation upon passage through a medium. The attenuation is caused by absorption and scattering.
Complex plane1	The complex plane contains the set of complex numbers. It is a plane spanned by the set of real numbers, normally along the horizontal axis, and the set of imaginary numbers, normally along the vertical axis.
Set1	In mathematics a set is a collection of related objects. The mathematical usage is similar to the ordinary English meaning of the word. The objects that make up a set are called the elements of the set. If a set contains an unlimited number of elements it is an infinite set. Otherwise it is a finite set.
Quantity1	A numerical value either scalar or vector, which describes some attribute of an object like its position or its velocity . We sometimes speak of physical quantities to signify that we are talking about an object's properties or attributes as opposed to a purely mathematical quantity.
Symmetry1	Symmetry refers to property that is now charged when operation or reference frame is charged.
Accuracy1	Accuracy refers to closeness of a measurement to the standard value of that quantity.
Vector1	Any quantity that has both magnitude and direction. Velocity is a vector.
Origin1	Origin refers to the point in a reference frame from which measurements are made. It is the location of the zero value for each axis in the frame.

Chapter 11. LINEAR FEEDBACK SYSTEMS

State1	Dynamical systems evolve over the course of time. The state of the system at any instant may be identified by the values of certain variables at that instant. For example specifying the angle from the vertical and the velocity of a frictionless pendulum allows us to predict its position and velocity at any future time. Therefore the state of the pendulum at any instant is its position and velocity. In this example the position and velocity are known as state variables.
Reflection1	Reflection refers to the change when light, sound, or other waves bounce backwards off a boundary .
Revolution1	Revolution refers to the orbital motion of one object around another. The Earth revolves around the Sun in one year. The moon revolves around the Earth in approximately 28 days.
Oscillation1	A to-and-fro or, side-to-side movement is an oscillation.
Nominal1	A nominal dimension is one that gives the intended or approximate size but this may (and often does) vary from the actual dimension. For example, a common lumber shape is a 2x4 but this is a nominal size and the actual dimensions are 1.5"e; x 3.5"e;. The word is from Latin, of a name, nomin-, nomen name thus can be thought of as 'what we call it'.
Precision1	Degree of exactness in a measurement is precision.
Natural frequency1	The frequency of vibration of an elastic object that depends on the size, composition, and shape of the object is referred to as natural frequency.
9. Thus a gev is a billion electron Volts. The mass-energy of a proton or neutron is approximately 1 gev.	The units one uses should be of a size that makes sense for the particular subject at hand. It is easiest to define units in each area of science and then relate them to one another than to go around measuring particle masses in grams or cheese in proton mass units. In particle physics the standard unit is the unit of energy gev. One ev is the amount of energy that an electron gains when it moves through a potential difference of 1 Volt . G stands for Giga, or 10
Rad1	One rad is equal to an energy absorption of 100 ergs in a gram of any material. An 'erg' is a unit for quantifying energy .
Gas1	State of matter that expands to fill container is referred to as gas.
Converge1	If the dependent variable of a function under iteration gets closer and closer to a fixed value the function is said to converge under iteration.

Chapter 11. LINEAR FEEDBACK SYSTEMS

Singularity1	The center of a black hole, where the curvature of spacetime is maximal is its singularity. At the singularity, the gravitational tides diverge; no solid object can even theoretically survive hitting the singularity. Although singularities generally predict inconsistencies in theory, singularities within black holes do not necessarily imply that general relativity is incomplete so long as singularities are always surrounded by event horizons. A proper formulation of quantum gravity may well avoid the classical singularity at the centers of black holes.
Gravity1	A force with infinite range which acts between objects, such as planets, according to their mass is called gravity.
Charge1	A property of atomic particles. Electrons and protons have opposite charges and attract each other. The charge on an electron is negative and that on a proton is positive. Measured in coulombs.
Electron1	Subatomic particle of small mass and negative charge found in every atom is called the electron.
Volt1	The unit of voltage or potential difference is a volt.
Resistor1	Device designed to have a specific resistance is a resistor.
Resistance1	Ratio of potential difference across device to current through it are called the resistance.
Capacitance1	Ratio of charge stored per increase in potential difference is called capacitance.
Acceleration1	Change in velocity divided by time interval over which it occurred is an acceleration.
Reference point1	Zero location in a coordinate system or frame of reference is a reference point.
Angular acceleration1	The rate of change of angular velocity of a body moving along a circular path is called its angular acceleration.
Equilibrium1	Equilibrium refers to condition in which net force is equal to zero. Condition in which net torque on object is zero.

PRACTICE QUIZ
Chapter 11. LINEAR FEEDBACK SYSTEMS

1. _____ refers to source of diverging light rays; either luminous or illuminated.

    a.
    b.
    c. Object
    d.

2. The rate of change of angular velocity of a body moving along a circular path is called its _____.

    a.
    b. Angular acceleration
    c.
    d.

3. _____ refers to potential difference. It is a measure of the change in energy that one coulomb of electric charge undergoes when moved between 2 points.

    a.
    b. Voltage
    c.
    d.

4. An _____ is a mathematical expression with an equal sign in it. It signifies that the numerical or vector value on one side of the = is the same as the numerical or vector value on the other side. An _____ may include variables and parameters. If any of the variables are rates of change, the _____ is called a differential equation.

    a. Equation
    b.
    c.
    d.

5. Defined collection of objects is called a _____.

    a.
    b. System
    c.
    d.

ANSWER KEY
Chapter 11. LINEAR FEEDBACK SYSTEMS

1. c
2. b
3. b
4. a
5. b

You can take the complete Chapter Practice Test

for Chapter 11. LINEAR FEEDBACK SYSTEMS
on all key terms, persons, places, and concepts.

Online 99 Cents

http://www.epub115.4.2097.11.cram101.com/

Use www.Cram101.com for all your study needs

including Cram101's online interactive problem solving labs in chemistry, statistics, mathematics, and more.

Other Cram101 e-Books and Tests

Want More?
Cram101.com...

Cram101.com provides the outlines and highlights of your textbooks, just like this e-StudyGuide, but also gives you the PRACTICE TESTS, and other exclusive study tools for all of your textbooks.

Learn More. *Just click*
http://www.cram101.com/